A Level
Mathematics
for Edexcel

Further
Pure

FP2

Mark Rowland

OXFORD
UNIVERSITY PRESS

Great Clarendon Street, Oxford OX2 6DP

Oxford University Press is a department of the University of Oxford.
It furthers the University's objective of excellence in research, scholarship,
and education by publishing worldwide in

Oxford   New York

Auckland   Cape Town   Dar es Salaam   Hong Kong   Karachi
Kuala Lumpur   Madrid   Melbourne   Mexico City   Nairobi
New Delhi   Shanghai   Taipei   Toronto

With offices in

Argentina   Austria   Brazil   Chile   Czech Republic   France   Greece
Guatemala   Hungary   Italy   Japan   South Korea   Poland   Portugal
Singapore   Switzerland   Thailand   Turkey   Ukraine   Vietnam

Oxford is a registered trade mark of Oxford University Press
in the UK and in certain other countries

British Library Cataloguing in Publication Data

Data available

ISBN 978-0-19-911788-8
10 9 8 7 6 5 4 3 2

Printed in Great Britain by Ashford Colour Press Ltd

Paper used in the production of this book is a natural,
recyclable product made from wood grown in sustainable forests.
The manufacturing process conforms to the environmental
regulations of the country of origin.

Acknowledgements

The photograph on the cover is reproduced courtesy of Rob Casey/Brand X/Corbis.

The Publisher would like to thank the following for permission to reproduce photographs:

p20 Mark Yuill/Shutterstock; **p28**Pictorial Press Ltd/Alamy; **p70**Mark Bond/Shutterstock; **p94** IT Stock/OUP;
p118 Mtoome/Dreamstime; **p142** Matthias Straka/Dreamstime; **p172**Grynold/Dreamstime.

Series managing editor Anna Cox

About this book

Endorsed by Edexcel, this book is designed to help you achieve your best possible grade in Edexcel GCE Further Mathematics Further Pure 2 unit.

Each chapter starts with a list of objectives and a 'Before you start' section to check that you are fully prepared. Chapters are structured into manageable sections, and there are certain features to look out for within each section:

Key points are highlighted in a blue panel.

Key words are highlighted in bold blue type.

Worked examples demonstrate the key skills and techniques you need to develop. These are shown in boxes and include prompts to guide you through the solutions.

Derivations and additional information are shown in a panel.

Helpful hints are included as blue margin notes and sometimes as blue type within the main text.

Misconceptions are shown in the right margin to help you avoid making common mistakes.

Investigational hints prompt you to explore a concept further.

Each section includes an exercise with progressive questions, starting with basic practice and developing in difficulty. Some exercises also include 'stretch and challenge' questions marked with a stretch symbol ▌·

At the end of each chapter there is a 'Review' section which includes exam style questions as well as past exam paper questions. There are also two 'Revision' sections per unit which contain questions spanning a range of topics to give you plenty of realistic exam practice.

The final page of each chapter gives a summary of the key points, fully cross-referenced to aid revision. Also, a 'Links' feature provides an engaging insight into how the mathematics you are studying is relevant to real life.

At the end of the book you will find full solutions, a key word glossary, a list of formulae given in the Edexcel formulae booklet and an index.

You can find a chapter to cover background knowledge for this unit on the A2 Teaching and Learning CD-ROM which accompanies this series.

Contents FP2

1

Inequalities

This chapter will show you how to

- solve inequalities involving rational linear expressions
- solve inequalities involving quadratic expressions
- solve equations and inequalities involving the modulus sign.

See Background knowledge Sections 0.1, 0.2 and 0.3 ◉.

Before you start

You should know how to:

1 Solve an inequality.

e.g. Solve the inequality $x^2 - 4x + 3 < 0$

$x^2 - 4x + 3 \equiv (x - 1)(x - 3)$

Sketch the graph of $y = (x - 1)(x - 3)$:

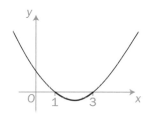

Values of y on the graph are negative for $1 < x < 3$.
The inequality $x^2 - 4x + 3 < 0$ has solution $1 < x < 3$.

2 Simplify an expression as a single fraction.

e.g. Simplify $\dfrac{x + 5}{x + 2} - 2$

$\dfrac{x + 5}{x + 2} - 2 \equiv \dfrac{(x + 5) - 2(x + 2)}{x + 2}$

$\equiv \dfrac{1 - x}{x + 2}$

3 Sketch the graph of a modulus function.

e.g. Sketch the graph of $y = |3x - 4|$

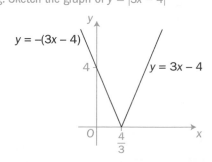

$y = -(3x - 4)$ $y = 3x - 4$

Check in:

See **C1** for revision.

1 Solve these inequalities.

a $x^2 - 7x - 8 < 0$

b $2x^2 + x - 1 > 0$

c $2x^2 \geqslant 5x + 3$

d $4x^2 \geqslant 12x - 9$

2 Simplify these expressions.   Refer to unit **C3**.

a $\dfrac{2x + 1}{x - 3} - 2$

b $\dfrac{5x + 1}{x - 1} + 2x$

c $\dfrac{2x + 1}{x + 1} - \dfrac{3x}{3x - 1}$

3 a On separate diagrams sketch these graphs.

   i $y = |3x + 2|$
   ii $y = |x^2 + 5x + 4|$

b Sketch, on one diagram, the graphs of $y = |2x + 1|$ and $y = |4 - 2x|$

FP2

A critical value of the rational function $\dfrac{f(x)}{g(x)}$ is a value of x for which either $f(x) = 0$ or $g(x) = 0$.

If $h(x) = \dfrac{f(x)}{g(x)}$, you can use the critical values of $h(x)$ to solve an inequality of the form $h(x) \leqslant 0$ or $h(x) \geqslant 0$.

EXAMPLE 1

Solve the inequality $\dfrac{2x-1}{x-2} < 0$

Let $h(x) = \dfrac{2x-1}{x-2}$

Find the critical values of $h(x)$:

The critical values of the function are $x = \dfrac{1}{2}$ and $x = 2$.

$2x - 1 = 0$ for $x = \dfrac{1}{2}$.
$x - 2 = 0$ for $x = 2$.

Use the critical values to divide the x-axis into three intervals.
Test the sign of $h(x)$ in each region:

$h\left(\dfrac{1}{2}\right) = 0$ $h(2)$ is undefined.

$h(x) > 0$	$h(x) < 0$	$h(x) > 0$
$x < \dfrac{1}{2}$	$\dfrac{1}{2}$ $\dfrac{1}{2} < x < 2$	2 $x > 2$

The intervals are called **critical regions**. A function does not change sign within each of its critical regions.

e.g. $h(0) = \dfrac{2 \times 0 - 1}{0 - 2}$ $h(1) = \dfrac{2 \times 1 - 1}{1 - 2}$ $h(3) = \dfrac{2 \times 3 - 1}{3 - 2}$

$= \dfrac{1}{2} (> 0)$ $= -1 (< 0)$ $= 5 (> 0)$

$h(x)$ is negative only in the interval $\dfrac{1}{2} < x < 2$.

The aim is to solve $h(x) < 0$.

Hence the inequality $\dfrac{2x-1}{x-2} < 0$ has solution $\dfrac{1}{2} < x < 2$.

You can also solve the inequality $\frac{2x-1}{x-2} < 0$ by considering *separately* the sign of $(2x-1)$ and $(x-2)$ over each critical region:

	$x < \frac{1}{2}$	$\frac{1}{2} < x < 2$	$x > 2$
$(2x-1)$	−	+	+
$(x-2)$	−	−	+	← $x - 2 > 0$ for all $x > 2$
$\frac{2x-1}{x-2}$	$\frac{(-)}{(-)} = +$	−	+

Each sign in the last row is determined by the other two signs in that column.

$\frac{2x-1}{x-2} < 0$ for $\frac{1}{2} < x < 2$

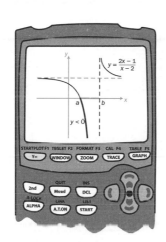

You can use the graph plotter on a calculator to check that the solution to $\frac{2x-1}{x-2} < 0$ is of the correct form. By plotting the graph with equation $y = \frac{2x-1}{x-2}$ you can see that values of x for which the graph lies *below* the x-axis (i.e. for which $y < 0$) lie between two numbers a and b.

EXAMPLE 2

Solve the inequality $\quad \frac{2x-3}{x+2} \geqslant 0$

If the inequality involves the $\geqslant$ or $\leqslant$ sign then the solution may include one or more of the critical values.

The critical values of $h(x) = \frac{2x-3}{x+2}$ are $x = \frac{3}{2}$ and $x = -2$.

Test the sign of $h(x)$ in each critical region:

The aim is to solve $h(x) \geqslant 0$.

$h(-2)$ is undefined. $\quad h\left(\frac{3}{2}\right) = 0$

```
        h(x) > 0  ┊   h(x) < 0  ┊   h(x) > 0
    ────────────────────────────────────────→ x
                 -2            3/2

          x < -2  ┊  -2 < x < 3/2  ┊  x > 3/2
```

e.g. $\quad h(-3) > 0 \qquad h(0) < 0 \qquad h(2) > 0$

The inequality $\frac{2x-3}{x+2} \geqslant 0$ has solution $x < -2$ or $x \geqslant \frac{3}{2}$.

You must include the critical value $x = \frac{3}{2}$ since $h\left(\frac{3}{2}\right) = 0$. Do not include $x = -2$ since $h(-2)$ is undefined.

Exercise 1.1

1 Solve each of these inequalities.

 a $\dfrac{x-3}{2x-1} > 0$ **b** $\dfrac{3x-1}{x-2} < 0$

 c $\dfrac{4x+5}{3x-5} \leqslant 0$ **d** $\dfrac{8-4x}{2x+3} > 0$

 e $\dfrac{5x-3}{3-x} \leqslant 0$ **f** $\dfrac{3-4x}{9-2x} > 0$

 g $\dfrac{5x}{3-2x} \geqslant 0$

2 Solve each of these inequalities.

 a $\dfrac{6x-1}{3-2x} \geqslant 0$ **b** $\dfrac{3-2x}{x} \leqslant 0$

3 By finding the critical value of each function, or otherwise, solve these inequalities.

 a $\dfrac{1}{x-3} > 0$ **b** $\dfrac{3}{2x-1} < 0$

 c $\dfrac{1}{3x+2} < 0$ **d** $\dfrac{2}{3-2x} > 0$

4 **a** Solve each inequality.

  i $\dfrac{x}{2x-3} \leqslant 0$ **ii** $\dfrac{2x-1}{x-2} \geqslant 0$

 b Hence write down the range of values of x which satisfy both inequalities.

5 **a** Solve each of these inequalities.

  i $\dfrac{3x-2}{2-x} > 0$ **ii** $\dfrac{5x-3}{x-3} \geqslant 0$

 b Hence show that no value of x satisfies both inequalities.

6 **a** Solve the inequality $\dfrac{x-2a}{x-4a} < 0$, giving your answer in terms of the positive constant a.

 b Given that the solution of this inequality is $1 < x < 2$, state the value of a.

7 a Find the range of values of x which satisfy both

$$\frac{4x-1}{3x+1} \geqslant 0 \quad \text{and} \quad \frac{1-3x}{4x+1} \geqslant 0$$

b Find the smallest positive value of the constant a such that no value of x satisfies both

$$\frac{4x-1}{3x+1} \geqslant 0 \quad \text{and} \quad \frac{x}{ax-1} \leqslant 0$$

8 Solve these inequalities.

a $\dfrac{x^2+3x}{x^2-2x} > 0$ **b** $\dfrac{2x^2+x-1}{x^2-1} \leqslant 0$

c $\dfrac{x^3+3x^2+2x}{4x-x^3} \geqslant 0$

9 Consider the inequality $\dfrac{2x-5}{k-2x} > 0$ where k is a real number.

a Solve this inequality when $k = -1$.

b Show that when $k = 5$ this inequality is not satisfied by any value of $x \neq 2.5$.

10 Solve these inequalities, giving each answer in terms of the constant $k > 0$.

a $\dfrac{x-k}{x+k} > 0$ **b** $\dfrac{2x-k}{x-k} < 0$

c $\dfrac{2x+k}{x+2k} < 0$ **d** $\dfrac{2k-x}{x-4k} > 0$

11 a Solve the inequality $\dfrac{k^2x-1}{kx-1} \geqslant 0$ giving your answer in terms of the constant $k > 1$.

b Solve the inequality $\dfrac{h^2x-1}{hx+1} < 0$ giving your answer in terms of the constant $h < -1$.

12 a For any pair of linear functions $f(x)$ and $g(x)$, prove that the inequality $\dfrac{g(x)}{f(x)} > 0$ has the same solution as the inequality $\dfrac{f(x)}{g(x)} > 0$.

b Give an example of a pair of linear functions $f(x)$ and $g(x)$ such that the inequalities $\dfrac{g(x)}{f(x)} \geqslant 0$ and $\dfrac{f(x)}{g(x)} \geqslant 0$ have different solutions.

Further rational linear inequalities

You can solve an inequality of the form $\frac{f(x)}{g(x)} \leqslant k$, for k a non-zero constant, by rearranging it into the form $\frac{f(x)}{g(x)} - k \leqslant 0$

Use the techniques from Section 1.1 to solve the rearranged inequality.

EXAMPLE 1

Find the values of x for which $\frac{2x+1}{x+2} \leqslant 1$

Rearrange the given inequality into an appropriate form:

$$\frac{2x+1}{x+2} \leqslant 1 \quad \text{so} \quad \frac{2x+1}{x+2} - 1 \leqslant 0$$

Do not multiply through by $(x+2)$ since it is negative for some values of x.

Simplify and solve the rearranged inequality:

$$\frac{2x+1}{x+2} - 1 \leqslant 0 \quad \text{so} \quad \frac{x-1}{x+2} \leqslant 0$$

$$\frac{2x+1}{x+2} - 1 = \frac{2x+1-x-2}{x+2} = \frac{x-1}{x+2}$$

The critical values of $h(x) = \frac{x-1}{x+2}$ are $x = 1$ and $x = -2$.

$h(-2)$ is undefined. $h(1) = 0$

Test the sign of $h(x)$ for values of x in each region.

$$
\begin{array}{ccc}
h(x) > 0 & h(x) < 0 & h(x) > 0 \\
\hline
\multicolumn{1}{c}{} & -2 \qquad\qquad 1 & \to x \\
x < -2 & -2 < x < 1 & x > 1
\end{array}
$$

$h(x) \leqslant 0$ for $-2 < x \leqslant 1$.

You must include $x = 1$ in the solution, since $h(1) = 0$.

Hence the inequality $\frac{2x+1}{x+2} \leqslant 1$ has solution $-2 < x \leqslant 1$.

The solution of $\frac{2x+1}{x+2} \leqslant 1$ is the same as the solution of $\frac{x-1}{x+2} \leqslant 0$

Exercise 1.2

1 Solve each of these inequalities.

    a $\dfrac{2x+3}{x+1} < 1$ **b** $\dfrac{4x+3}{x+2} > 2$ **c** $\dfrac{3+2x}{x-1} < 3$

    d $\dfrac{2x+6}{x+6} \leqslant -1$ **e** $\dfrac{5-2x}{2x-1} > -2$ **f** $\dfrac{1+2x}{3x} \leqslant \dfrac{1}{2}$

    g $\dfrac{4x}{2+3x} \leqslant \dfrac{1}{3}$ **h** $\dfrac{3-4x}{3-2x} \geqslant -1$

2 Find the set of values of x for which

    a $\dfrac{x+1}{x-2} > 1$ **b** $\dfrac{4x-1}{2x+1} < 2$ **c** $\dfrac{2x-1}{3x-1} < \dfrac{2}{3}$

3 Tom is trying to solve the inequality $\dfrac{1-2x}{3-x} > 2$

His working is shown:

Inequality to solve $\qquad\qquad\qquad \dfrac{1-2x}{3-x} > 2$

Step 1   Multiply both sides by $(3-x)$: $1 - 2x > 2(3-x)$

Step 2   Multiply out bracket: $1 - 2x > 6 - 2x$

Step 3   Subtract $1 - 2x$ from both sides: $0 > 5$

Since 0 is not greater than 5, no values of x satisfy the inequality.

Explain the error in his working, and give the correct solution.

4   **a**   Solve the inequalities   **i** $\dfrac{5x+3}{2x-3} \leqslant 1$ **ii** $\dfrac{4x-1}{2-x} \geqslant 2$

   b   Hence write down the range of values of x which satisfy both inequalities.

5   The integer x is such that $\dfrac{x}{x+1} > 1.024$

Find the least possible value of x.

6   **a**   Solve the inequality $\dfrac{x+a}{x-2} < 0$

Give your answer in terms of the positive constant a.

   b   Hence find, in terms of $a > 0$, the solution of each of these.

    i $\dfrac{2x+a-2}{x-2} < 1$ **ii** $\dfrac{4-2a-3x}{2-x} > 2$

7   Solve each of these inequalities.

You may use a calculator if you wish, but you should give each solution in simplified surd form.

   a $\dfrac{x}{\sqrt{2x}-1} > \sqrt{2}$ **b** $\dfrac{x}{\sqrt{2}-x} \leqslant 2$ **c** $\dfrac{x}{\sqrt{2}-x} \geqslant \sqrt{2}$

8   Solve these inequalities.

   a $\dfrac{x^2-3x}{3x^2-4x} > 1$ **b** $\dfrac{2x^2+3x+1}{x^2-1} < 2$ **c** $\dfrac{3-2x-5x^2}{x^2+4x+3} \leqslant \dfrac{1}{2}$

FP2

You can use critical values to solve an inequality involving quadratic expressions.

EXAMPLE 1

Solve the inequality $\dfrac{2x-1}{(x+1)(x-2)} \leqslant 0$

Find the critical regions and solve the inequality:

The critical values of $h(x) = \dfrac{2x-1}{(x+1)(x-2)}$ are $x=-1$, $x=\dfrac{1}{2}$ and $x=2$.

Identify the sign of h(x) in the four critical regions:

h(−1) is undefined. $h\left(\dfrac{1}{2}\right) = 0$ h(2) is undefined.   Test values of x in each interval.

| h(x) < 0 | h(x) > 0 | h(x) < 0 | h(x) > 0 |

$-1 \qquad \dfrac{1}{2} \qquad 2$

$x < -1 \quad -1 < x < \dfrac{1}{2} \quad \dfrac{1}{2} < x < 2 \quad x > 2$

e.g.   h(−2) < 0   h(0) > 0   h(1) < 0   h(3) > 0

The inequality $\dfrac{2x-1}{(x+1)(x-2)} \leqslant 0$

has solution $x < -1$ or $\dfrac{1}{2} \leqslant x < 2$.

You must include $x = \dfrac{1}{2}$

since $h\left(\dfrac{1}{2}\right) = 0$.

You can solve an inequality of the form f(x) > g(x) by solving h(x) > 0 where h(x) = f(x) − g(x)

EXAMPLE 2

Find the set of values of x for which $\dfrac{1}{x+1} > \dfrac{1}{x+3}$

Rearrange and simplify the inequality into the form h(x) > 0:

$\dfrac{1}{x+1} > \dfrac{1}{x+3}$ so $\dfrac{1}{x+1} - \dfrac{1}{x+3} > 0$ i.e. $\dfrac{2}{(x+1)(x+3)} > 0$

The critical values of $h(x) = \dfrac{2}{(x+1)(x+3)}$ are $x=-1$ and $x=-3$.

Identify the sign of h(x) in each of the critical regions:

h(x) > 0 for either $x < -3$ or $x > -1$.

Hence the inequality $\dfrac{1}{x+1} > \dfrac{1}{x+3}$

has solution $x < -3$ or $x > -1$.

h(−3) undefined.   h(−1) undefined.

| h(x) > 0 | h(x) < 0 | h(x) > 0 |

$-3 \qquad\qquad -1 \qquad\to x$

$x < -3 \quad -3 < x < -1 \quad x > -1$

e.g. h(−4) > 0   h(−2) < 0   h(0) > 0

Exercise 1.3

1 Solve these inequalities.

a $\dfrac{5x-4}{(x+1)(x-2)} > 0$

b $\dfrac{4x+5}{(2x-1)(x-3)} < 0$

c $\dfrac{2x+1}{(2x-1)(x+2)} \geqslant 0$

d $\dfrac{(5-2x)(3x-1)}{x+3} \leqslant 0$

e $\dfrac{x}{(x-2)(x+3)} > 0$

f $\dfrac{(x+3)(x+1)}{(x-1)(x-3)} \geqslant 0$

2 Solve these inequalities.

a $\dfrac{x+3}{x^2-5x+4} > 0$

b $\dfrac{2x-1}{x^2+6x+8} > 0$

c $\dfrac{2x^2+x-1}{x-2} < 0$

d $\dfrac{x+1}{2x^2+5x-3} \leqslant 0$

e $\dfrac{x^2-7x+10}{x^2-9} > 0$

3 Find the set of values of x which satisfy these inequalities.

a $\dfrac{x-2}{x^2+1} < 0$

b $\dfrac{x^2+3}{x^2+x-2} < 0$

c $\dfrac{x^2+9}{x^2-9} > 0$

4 Solve these inequalities.

a $\dfrac{3}{x-1} < \dfrac{2}{x}$

b $\dfrac{1}{x+3} > \dfrac{1}{2x+3}$

c $\dfrac{1}{3-x} \leqslant \dfrac{2}{2x+1}$

d $\dfrac{1}{4x-1} \geqslant \dfrac{1}{3x+1}$

e $\dfrac{x}{x-2} < \dfrac{x+3}{x-6}$

f $\dfrac{x+1}{x+3} \leqslant \dfrac{x}{x-1}$

5 Find the set of values of x which satisfy these inequalities.

a $\dfrac{3x}{2-x} > x$

b $\dfrac{10x}{3-x} < 2x+1$

c $\dfrac{2x-5}{x-1} \leqslant 1-x$

d $\dfrac{2-x}{2x+3} \geqslant 1-2x$

e $\dfrac{2x^2}{x+1} < x$

f $\dfrac{x^2-7}{x-1} \geqslant 4-x$

g $\dfrac{5}{x^2-4} \leqslant 1$

h $\dfrac{3x-1}{9-x^2} \leqslant 1$

i $\dfrac{4x+1}{x} > -4x$

6 Solve each inequality, giving your solutions in terms of the constant $k > 1$.

a $\dfrac{x^2-k^2}{x^2+k^2} < 0$

b $\dfrac{x^2-1}{x^2-k^2} > 0$

c $\dfrac{x^2+k^2}{x^2-k^2} > 1$

d $\dfrac{1}{1-kx} \geqslant kx+1$

FP2

Solving equations involving the modulus symbol

You can use a graphical method to solve an equation which involves the modulus symbol.

Solve the equation $|2x - 1| = 3$

On the same diagram sketch the graphs of the equations
$y = |2x - 1|$ and $y = 3$:

See **C3** for revision.

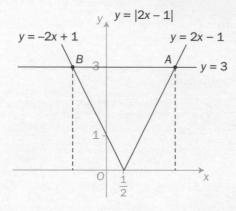

The graph of $y = |2x - 1|$ has y-intercept 1. This helps you correctly position the line $y = 3$

The solution of the equation $|2x - 1| = 3$ is given by the values of x at A and B where the two graphs intersect. See **C1** for revision.

Find the x-coordinates of the points A, B where the two graphs intersect to solve the given equation:

At A, $2x - 1 = 3$ so $x = 2$.
Intersection point A has x-coordinate $x = 2$.

At B, $-2x + 1 = 3$ so $x = -1$.
Intersection point B has x-coordinate $x = -1$.

Hence the equation $|2x - 1| = 3$ has solution $x = 2$ or $x = -1$.

You can also solve a modulus equation using algebra by replacing the modulus symbol with the $\pm$ symbol.

Use algebra to solve the equation $|4x - 3| = 5$

See **C3** for revision.

Replace the modulus symbol $|\ |$ with the $\pm$ symbol:
$|4x - 3| = 5$ so $\pm(4x - 3) = 5$

Solve the (+) and (−) equations separately:

(+) $4x - 3 = 5$ gives $4x = 8$, i.e. $x = 2$
(−) $-(4x - 3) = 5$ so $x = -\dfrac{1}{2}$

The (+) equation is $+(4x - 3) = 5$
The (−) equation is $-(4x - 3) = 5$

Hence the equation $|4x - 3| = 5$ has solution $x = 2$, $x = -\dfrac{1}{2}$.

▌Take care with the negative signs.

FP2

You can use algebra to solve an equation of the form $|f(x)| = g(x)$, but you should check that each answer is valid.

Check your answer by using substitution or a graphical method.

EXAMPLE 3

a Use algebra to solve the equation $|x - 1| = 2x + 3$

b Sketch a graph to illustrate the solution of this equation.

---

a Replace the modulus symbol $|\ |$ with the $\pm$ symbol:
$$|x - 1| = 2x + 3 \quad \text{so} \quad \pm(x - 1) = 2x + 3$$

Solve the $(+)$ and $(-)$ equations separately:
$$(+) \quad x - 1 = 2x + 3 \quad \text{so} \quad x = -4$$

Substitute this answer into the given equation to check its validity:
$$|x - 1| = 2x + 3$$

$x = -4$ so LHS $= |-4 - 1| = 5$ and RHS $= 2 \times (-4) + 3 = -5$
Hence $x = -4$ is *not* a valid answer.

The LHS and RHS are not equal when $x = -4$.

$$(-) \quad -(x - 1) = 2x + 3 \quad \text{so} \quad x = -\frac{2}{3}$$
$$|x - 1| = 2x + 3$$

$$x = -\frac{2}{3} \quad \text{so} \quad \text{LHS} = \left|-\frac{2}{3} - 1\right| = \frac{5}{3}$$

$$\text{and} \quad \text{RHS} = 2 \times \left(-\frac{2}{3}\right) + 3 = \frac{5}{3}$$

The LHS and RHS are equal when $x = -\frac{2}{3}$.

i.e. $x = -\frac{2}{3}$ is a valid answer.

Hence the equation $|x - 1| = 2x + 3$ has solution $x = -\frac{2}{3}$.

b Sketch, on the same diagram, the graphs with equations $y = |x - 1|$ and $y = 2x + 3$:

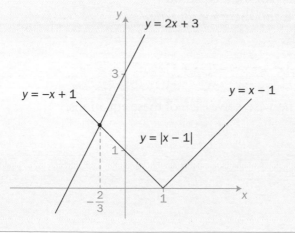

By comparing gradients, the line $y = 2x + 3$ is steeper than the line $y = x - 1$, as shown in the sketch.

If the line $y = x - 1$ were extended below the x-axis it would intersect the line $y = 2x + 3$ at $x = -4$.

FP2

Exercise 1.4

1 Use algebra to solve each equation.
   Check the validity of each answer by substitution.

   a $|2x - 1| = 5$ **b** $|4 - 3x| = 7$

   c $\left|\frac{1}{2}x + 1\right| = 4$ **d** $|3x - 2| = x$

   e $|2x - 5| = x + 1$ **f** $|4x + 1| = 7 - 2x$

2 Use algebra to solve each of these equations.
   Use a sketch to check the validity of each answer.

   a $|x^2 - 1| = 3$ **b** $|x^2 - 10| = 6$

   c $|x^2 - 4x| = 12$

3 **a** Sketch, on the same diagram, the graphs with equation

       i $y = |x^2 - 4|$ **ii** $y = 2x + 11$

       Label the points where each graph crosses the coordinate axes.

   b Hence, or otherwise, solve the equation

       $|x^2 - 4| = 2x + 11$

4 Use an appropriate method to solve each of these equations.

   a $|x^2 - 2x| = x + 4$ **b** $|x^2 - 9| = 8x$

   c $|9x^2 - 1| = 5 - 3x$ **d** $|4x^2 - 9| = 2x + 3$

   e $|x^2 - x - 2| = x + 1$

5 Use a graphical method to solve each of these equations,
   giving answers in simplified surd form where appropriate.

   a $|x^2 - 6| = 2$ **b** $|x^2 - 2x| = 1$

   c $|x^2 - 4x + 3| = 2x - 4$ **d** $|4x^2 + 5x| = -2(2x + 1)$

6 Using a graphical approach, or otherwise, solve each of these equations.
   Check the validity of each answer.

   a $|x + 2| = |2x - 1|$

   b $|2x - 3| = |4x + 1|$

   c $|3x + 2| = |4 + 3x|$

7   Use a graphical method to solve each of these equations.
    Give your answers in simplified surd form where appropriate.

a $|x^2 - 9| = |4x + 3|$

b $|x^2 - 4| = |4x + 1|$

c $|x^2 - 3x| = |3x - 2|$

d $|x^2 - x - 6| = x + 6$

e $|2x^2 - 3x - 2| = 3(x + 1)$

f $|3 - 4x - 4x^2| = 2(1 - x)$

8   Show that the equation $|x^2 + 2x| = 3x - 6$ has no real solutions.

9   It is given that exactly three distinct values of x satisfy the
    equation $|x^2 - 8| = a$, where a is a positive constant.
    Find the value of a and hence solve the equation $|x^2 - 8| = a$

10 Solve these equations, giving answers in terms of the positive
    constant k.

a $|x^2 - k^2| = k^2 + 1$

b $2|x^2 - k^2| = k^2$

c $|x^2 - k^2| = 2kx + k^2$

You should use a graphical method to solve an inequality involving the modulus symbol. This helps you identify any values of x which are not valid answers for the inequality.

EXAMPLE 1

Find the set of values of x for which $|3x - 1| < x + 1$

Sketch, on the same diagram, the graphs with equations $y = |3x - 1|$ and $y = x + 1$:

See Section 1.4.

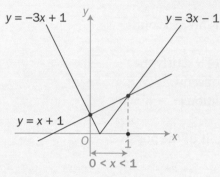

$y = -3x + 1$ $y = 3x - 1$

$y = x + 1$

$0 < x < 1$

Find the values of x where the graphs intersect by solving appropriate equations:

$(+)$ $3x - 1 = x + 1$ so $2x = 2$

             i.e. $x = 1$

$(-)$ $-3x + 1 = x + 1$ so $-4x = 0$

             i.e. $x = 0$

Inspect the diagram and look for appropriate sections which describe the given inequality:

For a given value of x in the solution of $|3x - 1| < x + 1$, the y-coordinate of the graph $y = |3x - 1|$ must be less than the corresponding y-coordinate of the graph $y = x + 1$

Hence the inequality $|3x - 1| < x + 1$ is satisfied by the values of x for which the graph of $y = |3x - 1|$ lies strictly *below* the graph of $y = x + 1$

The inequality $|3x - 1| < x + 1$ has solution $0 < x < 1$.

$x = 0$ and $x = 1$ are not included in the solution, as the question involved the $<$ symbol.

You can solve inequalities involving the modulus of a
quadratic function.

EXAMPLE 2

Find the complete set of values of x for which $|x^2 - 4| \geqslant 3x$

Sketch on the same diagram the graphs with equations
$y = |x^2 - 4|$ and $y = 3x$:

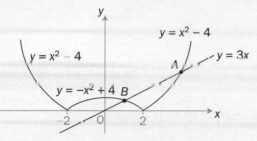

Find the x-coordinates of the intersection points A and B to
solve the inequality:

At intersection point A $\qquad x^2 - 4 = 3x$

i.e. $\qquad x^2 - 3x - 4 = 0$

$\qquad\qquad (x + 1)(x - 4) = 0$

Hence $x = 4$.

At intersection point B $\qquad -x^2 + 4 = 3x$

i.e. $\qquad x^2 + 3x - 4 = 0$

$\qquad\qquad (x - 1)(x + 4) = 0$

Hence $x = 1$.

The inequality $|x^2 - 4| \geqslant 3x$ is satisfied by the values of x
for which the graph of $y = |x^2 - 4|$ lies *on or above* the
graph of $y = 3x$

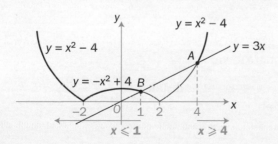

Hence the inequality $|x^2 - 4| \geqslant 3x$ has solution $x \leqslant 1$ or $x \geqslant 4$.

Ignore $x = -1$ since this does not
correspond to an intersection
point of the graphs.

Ignore the value $x = -4$.

FP2

Exercise 1.5

When not exact, give answers in simplified surd form.

1 Solve each of these inequalities.

    a $|2x - 1| < 5$

    b $|3x - 2| > 4$

    c $|8x + 3| \leqslant 1$

    d $|3x - 4| > x + 2$

    e $|2x + 3| \leqslant 3x + 1$

    f $|1 - 4x| > 2x - 1$

2 Find the set of values of x which satisfies each of these inequalities.

    a $|x^2 - 4| > 5$

    b $|x^2 - 5| < 4$

    c $|x^2 - 9| \geqslant x + 3$

    d $|x^2 - 6| \leqslant 6 - x$

    e $|x^2 - 2x| > x$

    f $|2x^2 - 7x + 3| > 3$

    g $|x^2 - 2x - 3| < 3 - x$

3 Solve each of these inequalities.

    a $|x^2 - 9| < 5$

    b $|4x^2 - 9| \geqslant 7$

    c $|x^2 - x - 6| < 1 - 3x$

    d $|x^2 + 3x - 4| \leqslant 2 - x$

4 **a** Express $x^2 + 4x - 1$ in completed square form.

    b Sketch the graph of the equation $y = x^2 + 4x - 1$ and state the coordinates of vertex of this graph.

    c Solve the inequality $|x^2 + 4x - 1| \leqslant 4$

5 Find the complete set of values of x for which

    a $|x^2 + 2x - 2| < 2 - x$

    b $|x^2 - 4x + 1| > 1$

    c $|x^2 + 6x + 7| \geqslant 9 - 4x$

FP2

6 Solve each of these inequalities.

a $|x - 2| < |3x + 2|$

b $|2x - 1| > |2x + 3|$

c $|x^2 - 6| > |x|$

d $|x^2 + 3x| \leqslant |x + 1|$

e $|x^2 - 6x + 5| \geqslant |2x - 3|$

7 Use a graphical method to solve the inequality $|2x - 1| \leqslant \dfrac{1}{x}$

8 The diagram shows the graph with equation $y = \dfrac{x - 1}{x - 2}$ for $x \neq 2$.

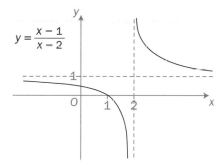

a Sketch, on a new diagram, the graph with equation $y = \left|\dfrac{x - 1}{x - 2}\right|$

b Hence, or otherwise, solve

   i $\left|\dfrac{x - 1}{x - 2}\right| \leqslant 2$

   ii $|x - 1| > |x - 2|$

   iii $\left|\dfrac{x - 1}{x - 2}\right| < x$

   Give answers in surd form where appropriate.

1 a Sketch, on the same diagram, the graph with equation $y = |2x^2 - 13|$ and the line with equation $y = 5$

b Hence, or otherwise, solve the inequality $|2x^2 - 13| \leqslant 5$

2 Find the range of values of x for which $\dfrac{5x}{2x - 1} > 3$

3 Find the complete set of values of x for which $\dfrac{x}{x + 1} > \dfrac{x - 1}{x}$

4 Determine all *integers* x which satisfy the inequality $\dfrac{x + 10}{7 - 2x} > 2$

5 a Sketch, on the same diagram, the graph with equation $y = |5x - 2|$ and the line with equation $y = 3x$
Find the x-coordinates of the points where these graphs intersect.

b Hence solve the inequality $|5x - 2| \geqslant 3x$

6 a Sketch, on the same diagram, the graphs with equations
$$y = |4x - 3| \quad \text{and} \quad y = |5 - x|$$

b Hence solve the inequality $|4x - 3| > |5 - x|$

7 Solve the inequality $|x^2 - 4| < 2x - 3$, giving your answer in simplified surd form.

8 a Sketch the graph with equation $y = |2x - a|$, given that $a > 0$.

b Hence solve the inequality $|2x - a| < x + 2a$, giving your answer in terms of a.

9 Solve the inequality $\dfrac{3x}{3x - 2} > \dfrac{x + 2}{x + 1}$

10 Solve the inequality $\dfrac{3x}{2x + k} < 1$. Give your answer in terms of $k > 0$.

11 a Sketch, on the same diagram, the graph with equation $y = |x^2 - 6x + 5|$ and the line with equation $y = 4 - 2x$

b Use algebra to find the solution of the equation $|x^2 - 6x + 5| = 4 - 2x$
Give your answers in simplified surd form.

c Solve the inequality $|x^2 - 6x + 5| \geqslant 4 - 2x$

12 Solve the inequality $\dfrac{x}{3x-2} < \dfrac{1}{x}$

13 Solve, by means of a sketch graph, the inequality $|2x^2 - x - 6| \geqslant |2x + 3|$

14 a Given that $a > 0$, sketch, on the same diagram, the graph with equation $y = |x^2 - a^2|$ and the line with equation $y = a^2 - 2ax$

  b Hence solve, in terms of a, the inequality $|x^2 - a^2| < a^2 - 2ax$
    Give your answers in surd form.

15 a Use algebra to solve the equation $|x(x - 2a)| = a(2a - x)$
    Give your answers in terms of the constant $a > 0$.

  b Sketch, on the same diagram, the graph with equation $y = |x(x - 2a)|$ and the line with equation $y = a(2a - x)$

  c Hence solve the inequality $|x(x - 2a)| \leqslant a(2a - x)$

16 a Solve the equation $|x + 1| = x + 3$

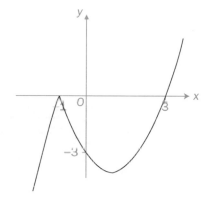

  The diagram shows a sketch of the curve $y = |x + 1|(x - 3)$

  b On a copy of the diagram, sketch the graph with equation $y = x^2 - 9$

  c Hence solve the inequality $|x + 1|(x - 3) < x^2 - 9$

17 a On the same diagram, sketch the curve with equation $y = |2x^2 + 5x - 3|$ and the line with equation $y = 3x + 3$

  b Use algebra to find the exact solutions of the equation $|2x^2 + 5x - 3| = 3x + 3$ giving answers in simplified surd form where appropriate.

  c Find the set of values of x for which $|2x^2 + 5x - 3| > 3x + 3$

FP2

Summary

Refer to

- The **critical values** of $\dfrac{f(x)}{g(x)}$ are values of x for which either
  $f(x) = 0$ or $g(x) = 0$.

  1.1–1.3

- You can find the solution of an inequality by investigating the sign of an appropriate function in each of its **critical regions**.

- To solve an inequality of the form $f(x) \leqslant g(x)$ (or $f(x) \geqslant g(x)$), rearrange it into the form $f(x) - g(x) \leqslant 0$ (or $f(x) - g(x) \geqslant 0$)

  1.4, 1.5

- You can solve an inequality involving the modulus symbol by using a graphical technique.

FP2

Links

Inequalities have many applications in the real world, especially within industry and business management.

Manufacturers can use inequalities to minimise waste during production of a particular item. They can also maximise profits by identifying the optimum quantities of the product to make as well as the most suitable price that the item should be sold at.

The area of mathematics which studies the use of inequalities in this way is called Linear Programming and is covered in the Decision Mathematics units.

2 Series

This chapter will show you how to
- use the method of differences to prove standard and other results for series
- efficiently evaluate series such as $\displaystyle\sum_{r=1}^{20}\frac{1}{r(r+1)}$

See Section 0.1

Before you start

You should know how to:

1 Factorise algebraic expressions.

e.g. Completely factorise $(2n+3)(n+2)-n^2$

$(2n+3)(n+2)-n^2 \equiv (2n^2+7n+6)-n^2$

$= n^2+7n+6$

$= (n+1)(n+6)$

2 Work with standard series results.

e.g. Simplify $\displaystyle\sum_{r=1}^{n}(4r-1)$

$\displaystyle\sum_{r=1}^{n}(4r-1) = 4\sum_{r=1}^{n}r - \sum_{r=1}^{n}1$

$= 4\left(\frac{1}{2}n(n+1)\right) - n$

$= n(2n+1)$

3 Add and subtract algebraic fractions.

e.g. Simplify $\dfrac{1}{n-1}+\dfrac{1}{n+1}$

$\dfrac{1}{n-1}+\dfrac{1}{n+1} \equiv \dfrac{(n+1)+(n-1)}{(n-1)(n+1)}$

$\equiv \dfrac{2n}{(n-1)(n+1)}$

4 Express a quotient in partial fractions.

e.g. Express $\dfrac{r+4}{r(r+1)}$ in partial fractions.

$\dfrac{r+4}{r(r+1)} \equiv \dfrac{A}{r}+\dfrac{B}{(r+1)}$

so $r+4 \equiv A(r+1)+Br$

$r=0$ gives $A=4$ and $r=-1$ gives $B=-3$

Hence $\dfrac{r+4}{r(r+1)} \equiv \dfrac{4}{r}-\dfrac{3}{(r+1)}$

Check in:

1 Factorise completely   See **C1** for revision.

  a n^3+4n^2+3n

  b $n(n+1)^2-4n$

  c $(n^2-1)(n+3)+3(n+1)$

  d $(3n+1)^2-(n-2)^2$

2 Use standard series to simplify

  a $\displaystyle\sum_{r=1}^{n}(2r+3)$ See **FP1** for revision.

  b $\displaystyle\sum_{r=1}^{n}(6r^2-1)$

  c $\displaystyle\sum_{r=1}^{n}4r(r^2-1)$

3 Simplify these expressions.

  a $2+\dfrac{1}{n+1}$

  b $1+\dfrac{1}{n}-\dfrac{1}{n+1}$

  c $\dfrac{n}{n+8}-\dfrac{2}{n+2}$

4 Write these expressions as partial fractions.

  a $\dfrac{3r}{(r+1)(r-2)}$ See **C4** for revision.

  b $\dfrac{2-r}{r^2-1}$

  c $\dfrac{4}{r(r+1)(r+2)}$

2.1　The method of differences

You can evaluate a certain type of series by grouping its terms in an appropriate way.

This process is known as the method of differences.

EXAMPLE 1

Use the method of differences to evaluate $\displaystyle\sum_{r=1}^{30}\left[(r+1)^4 - r^4\right]$

Write out the first few and last few terms of the series:

$$(2^4 - 1^4) + (3^4 - 2^4) + (4^4 - 3^4) + \cdots + (30^4 - 29^4) + (31^4 - 30^4)$$

Group the terms in pairs and write the sum vertically:

$$\sum_{r=1}^{30}(r+1)^4 - r^4 = (2^4 - 1^4)$$
$$+ (3^4 \quad - 2^4)$$
$$+ (4^4 \quad - 3^4)$$
$$+ \quad \cdots$$
$$+ (30^4 \quad - 29^4)$$
$$+ (31^4 \quad - 30^4)$$

Only the terms 31^4 and 1^4 survive the cancellation process.

After cancellation, the sum is $= 31^4 - 1^4 = 923\,520$

Hence $\displaystyle\sum_{r=1}^{30}\left(r+1\right)^4 - r^4 = 923\,520$

Remember to *subtract* 1^4 from 31^4.

You can use this method to evaluate summations involving algebraic fractions.

EXAMPLE 2

a Express $\dfrac{1}{r} - \dfrac{1}{(r+1)}$ as a single fraction.

b Hence find $\displaystyle\sum_{r=1}^{n}\frac{1}{r(r+1)}$

You may need to use partial fractions. See **C4** for revision.

Simplify your answer as far as possible.

a $\dfrac{1}{r} - \dfrac{1}{(r+1)} = \dfrac{(r+1) - r}{r(r+1)}$

$\qquad\qquad\qquad = \dfrac{1}{r(r+1)}$

b Use part **a** and the method of differences:

$$\sum_{r=1}^{n}\frac{1}{r(r+1)} = \sum_{r=1}^{n}\left[\frac{1}{r} - \frac{1}{r+1}\right]$$

The summation involves a subtraction so use the method of differences.

EXAMPLE 2 (CONT.)

Write out in vertical pairs the first few and last few terms of the right-hand side:

$$\sum_{r=1}^{n}\left[\frac{1}{r} - \frac{1}{r+1}\right] = \left(\frac{1}{1} - \frac{1}{2}\right)$$
$$+ \left(\frac{1}{2} - \frac{1}{3}\right)$$
$$+ \left(\frac{1}{3} - \frac{1}{4}\right)$$
$$+ \quad \cdots$$
$$+ \left(\frac{1}{n-1} - \frac{1}{n}\right)$$
$$+ \left(\frac{1}{n} - \frac{1}{n+1}\right)$$

Only the first and last terms survive the cancellation process.

Hence $\displaystyle\sum_{r=1}^{n}\frac{1}{r(r+1)} = 1 - \frac{1}{n+1} = \frac{n}{n+1}$

You may need to list several terms of a series before the cancellations become obvious.

EXAMPLE 3

a Simplify $(r+1)^2 - (r-1)^2$

b Hence, using the method of differences, show that

$$\sum_{r=1}^{n} r = \frac{1}{2}n(n+1)$$

---

a Expand the brackets and collect like terms:

$$(r+1)^2 - (r-1)^2 = (r^2 + 2r + 1) - (r^2 - 2r + 1)$$
$$= 4r$$

b $\displaystyle\sum_{r=1}^{n} 4r = \sum_{r=1}^{n}\left[(r+1)^2 - (r-1)^2\right]$

$$= (2^2 \quad - \quad 0^2)$$
$$+ (3^2 \quad - \quad 1^2)$$
$$+ (4^2 \quad - \quad 2^2)$$
$$+ (5^2 \quad - \quad 3^2)$$
$$+ \quad \cdots$$
$$+ ((n-1)^2 - (n-3)^2)$$
$$+ (n^2 \quad - (n-2)^2)$$
$$+ ((n+1)^2 - (n-1)^2)$$

You must use the method of differences, even though this sum can be found more easily by using a standard result.

The first cancellation occurs in the 1st and 3rd brackets: $2^2 - 2^2 = 0$
The second cancellation occurs in the 2nd and 4th brackets: $3^2 - 3^2 = 0$

Only the terms $(n+1)^2$, n^2, 1^2 and 0^2 survive the cancellation process.

Hence $\displaystyle\sum_{r=1}^{n} 4r = (n+1)^2 + n^2 - 1$

$$= 2n(n+1)$$

i.e. $\displaystyle\sum_{r=1}^{n} r = \frac{1}{2}n(n+1)$

$\displaystyle\sum_{r=1}^{n} r = \frac{1}{4}(2n(n+1)) = \frac{1}{2}n(n+1)$

Exercise 2.1

1 Use the method of differences to evaluate each of these series.

Where appropriate you may use without proof any standard results of summations.

   a $\quad 1 - 2 + 2 - 3 + 3 - 4 + \cdots + 20 - 21$

   b $\quad 3 - 2 + 4 - 3 + 5 - 4 + \cdots + 15 - 14$

   c $\quad 2 - 4 + 3 - 5 + 4 - 6 + 5 - 7 + \cdots + 40 - 42$

   d $\quad 4 - 1 + 5 - 2 + 6 - 3 + 7 - 4 + \cdots + 19 - 16$

2 **a** Simplify $r(r + 1) - r(r - 1)$

   b Hence use the method of differences to prove that

$$\sum_{r=1}^{n} r = \frac{1}{2} n(n + 1)$$

3 **a** Simplify $r^2 - (r - 1)^2$

   b Hence, using the method of differences, show that

$$\sum_{r=1}^{n} (2r - 1) = (n + 1)^2 \text{ for all positive integers } n.$$

   c Evaluate the series $\displaystyle\sum_{r=11}^{19} (2r - 1)$

4 **a** Find the value of the constant A such that
$$(2r + 1)^2 - (2r - 1)^2 = Ar$$
for all real values of r.

   b Hence use the method of differences to prove that

$$\sum_{r=1}^{n} r = \frac{1}{2} n(n + 1)$$

5 **a** Show that $\dfrac{1}{r + 1} - \dfrac{1}{r + 2} = \dfrac{1}{(r + 1)(r + 2)}$ for all real values of $r \geqslant 0$.

   b Hence, by using the method of differences, show that

$$\sum_{r=1}^{n} \frac{1}{(r + 1)(r + 2)} = \frac{n}{2(n + 2)}$$

   c Evaluate the series $\displaystyle\sum_{r=9}^{20} \frac{1}{(r + 1)(r + 2)}$

   Give your answer as a fraction in its simplest terms.

6 **a** Express $\dfrac{3}{(3r - 2)(3r + 1)}$ in partial fractions.

   b Hence find in terms of n an expression for

$$\sum_{r=1}^{n} \frac{1}{(3r - 2)(3r + 1)}$$

   c Deduce that $\displaystyle\sum_{r=1}^{n} \frac{1}{(3r - 2)(3r + 1)} < \frac{1}{3}$ for all $n > 1$.

7 a Express $\dfrac{6}{9r^2 - 12r - 5}$ in partial fractions.

b Hence show that $\displaystyle\sum_{r=1}^{n} \dfrac{4}{9r^2 - 12r - 5} = \dfrac{n(3n-5)}{(3n+1)(3n-2)}$

c Find the value of $\displaystyle\sum_{r=6}^{15} \dfrac{4}{9r^2 - 12r - 5}$

Give your answer to 3 significant figures.

8 a Use the method of differences to show that

$$\sum_{r=2}^{n} \ln\left(\frac{r+1}{r-1}\right) = \ln(1 + 2 + 3 + \cdots + n) \text{ for any } n \geqslant 2.$$

b Express, in a similar form, $\displaystyle\sum_{r=2}^{n} \ln\left(\dfrac{r^2 + 2r + 1}{r^2 - 2r + 1}\right)$

9 a Express $\dfrac{2}{r(r+1)(r+2)}$ in partial fractions.

b Hence show that $\displaystyle\sum_{r=1}^{n} \dfrac{4}{r(r+1)(r+2)} = \dfrac{n(n+3)}{(n+1)(n+2)}$

10 a Express $\dfrac{r^2 + r + 1}{r^2 + r}$ in the form $A + \dfrac{B}{r} + \dfrac{C}{r+1}$, for constants A, B and C to be determined.

b Hence show that $\displaystyle\sum_{r=1}^{n} \dfrac{r^2 + r + 1}{r^2 + r} = \dfrac{n(n+2)}{n+1}$

c Find an expression, in terms of n, for

$$\sum_{r=n-1}^{2n} \dfrac{r^2 + r + 1}{r^2 + r}, \quad \text{where } n \geqslant 2$$

Factorise your answer as far as possible.

11 a Express $(r+1)^3 - r^3$ in the form $Ar^2 + Br + 1$, for constants A and B to be found.

b Hence, using the method of differences, show that

$$\sum_{r=1}^{n} r^2 = \frac{1}{6}n(n+1)(2n+1)$$

12 Show that $(r+1)^4 - (r-1)^4 \equiv 8(r^3 + r)$

Hence use the method of differences to show that

$$\sum_{r=1}^{n} r^3 = \frac{1}{4}n^2(n+1)^2$$

Where appropriate you may use without proof any standard results of summations

1 a Simplify $(r+2)^2 - r^2$

b Hence, using the method of differences, show that
$$\sum_{r=1}^{n} r = \frac{1}{2}n(n+1)$$

2 a Express $\dfrac{4}{(4r-1)(4r+3)}$ in partial fractions.

b Hence, by using the method of differences, show that
$$\sum_{r=1}^{n} \frac{1}{(4r-1)(4r+3)} = \frac{n}{3(4n+3)}$$

c Evaluate the series $\displaystyle\sum_{r=16}^{32} \frac{1}{(4r-1)(4r+3)}$

Give your answer to 3 significant figures.

3 a Show that $(r+2)^3 - r^3 \equiv Ar^2 + Br + C$ for constants A, B and C to be stated.

b Hence, or otherwise, show that $\displaystyle\sum_{r=1}^{n} r^2 = \frac{1}{6}n(n+1)(2n+1)$ for all $n \geqslant 1$.

4 a Express $\dfrac{4}{r^2 - 1}$ in partial fractions.

b Hence show that $\displaystyle\sum_{r=2}^{n} \frac{4}{r^2 - 1} = \frac{(3n+2)(n-1)}{n(n+1)}$ for all $n > 2$.

c i Express $\dfrac{r^2+1}{r^2-1}$ in the form $P + \dfrac{Q}{r^2-1}$ for constants P and Q.

ii Hence find an expression for $\displaystyle\sum_{r=2}^{n}\left(\frac{r^2+1}{r^2-1}\right)$ in terms of n,

factorising your answer as far as possible.

5 a Show that $\dfrac{1}{r^2} - \dfrac{1}{(r+1)^2} \equiv \dfrac{2r+1}{r^2(r+1)^2}$

b Hence find an expression for $\displaystyle\sum_{r=1}^{n} \frac{2r+1}{r^2(r+1)^2}$ in the form $A + \dfrac{B}{(n+1)^2}$

for constants A and B to be stated.

c Evaluate the series $\displaystyle\sum_{r=4}^{19} \frac{2r+1}{r^2(r+1)^2}$

d Show that $\displaystyle\sum_{r=n}^{2n-1} \frac{2r+1}{r^2(r+1)^2} = \frac{3}{4n^2}$ for all positive integers n.

FP2

6 a Express $\dfrac{4}{(4r-3)(4r+1)}$ in partial fractions.

b Hence show that $\displaystyle\sum_{r=1}^{n}\left(4r+\dfrac{1}{(4r-3)(4r+1)}\right)=\dfrac{n(4n+3)(2n+1)}{4n+1}$

7 a Find the value of the constants A and B such that
$$(2r+1)^4-(2r-1)^4\equiv Ar^3+Br$$

b Hence show that $\displaystyle\sum_{r=1}^{n}r^3=\dfrac{1}{4}n^2(n+1)^2$

8 a Express $\dfrac{1}{4r^2+4r-3}$ in partial fractions.

b Hence show that $\displaystyle\sum_{r=1}^{n}\dfrac{3}{4r^2+4r-3}=\dfrac{n(4n+5)}{(2n+3)(2n+1)}$

9 a Show that $\dfrac{1}{r}-\dfrac{1}{(r+1)}+4=\dfrac{(2r+1)^2}{r(r+1)}$ for all $r>0$.

b Hence find, in terms of the positive integer n,
an expression for $\displaystyle\sum_{r=1}^{n}\dfrac{(2r+1)^2}{r(r+1)}$,
factorising your answer as far as possible.

c Evaluate the series $\displaystyle\sum_{r=6}^{14}\dfrac{(2r+1)^2}{r(r+1)}$

10 a Show that $\dfrac{r^3-r^2+1}{r^2-r}=r+\dfrac{1}{(r-1)}-\dfrac{1}{r}$ for all $r>1$.

b Hence find an expression for $\displaystyle\sum_{r=2}^{n}\left(\dfrac{r^3-r^2+1}{r^2-r}\right)$,
factorising your answer as far as possible.

11 a Express $\dfrac{2}{r(r^2-1)}$ in the form $\dfrac{A}{(r+1)}+\dfrac{B}{r}+\dfrac{C}{(r-1)}$ for constants A, B and C.

b Hence show that $\displaystyle\sum_{r=2}^{n}\dfrac{4}{r(r^2-1)}=\dfrac{(n-1)(n+2)}{n(n+1)}$

FP2

Exit →

Summary

Refer to

○ You can prove standard results for series using the method of differences. These include

2.1

$$\sum_{r=1}^{n} r = \frac{1}{2}n(n+1)$$

$$\sum_{r=1}^{n} r^2 = \frac{1}{6}n(n+1)(2n+1)$$

$$\sum_{r=1}^{n} r^3 = \frac{1}{4}n^2(n+1)^2$$

○ To find a sum of terms involving a quotient, use partial fractions.

2.1

Links

Sequences and series can be used in the financial world to model the long-term value of an investment or to predict the length of time needed to pay back a loan or a mortgage.

E.g. The total interest earned on £1000 invested at a compound interest rate of 5% over n years is given by the geometric series $\sum_{r=1}^{n} 50 \times 1.05^{r-1}$

FP2

28

3

Further complex numbers

This chapter will show you how to

- express a complex number in exponential form
- solve equations such as $z^5 = 1$ completely
- apply de Moivre's theorem to raise complex numbers to large powers
- prove trigonometric identities
- describe and transform loci such as straight lines and circles in the complex plane.

Refer to **FP1** Chapter 2 for revision.

See Section 0.7 .

FP2

Before you start

You should know how to:

1 Express a complex number in modulus–argument form.

e.g. Express $3 + i\sqrt{3}$ in modulus–argument form.

If $z = 3 + i\sqrt{3}$ then $|z| = \sqrt{9 + 3} = \sqrt{12}$

and $\arg z = \tan^{-1}\left(\dfrac{\sqrt{3}}{3}\right) = \dfrac{1}{6}\pi$

Hence $z = \sqrt{12}\left(\cos\left(\dfrac{1}{6}\pi\right) + i\sin\left(\dfrac{1}{6}\pi\right)\right)$

2 Find the equation of the perpendicular bisector of a line between two points.

e.g. Find an equation for the perpendicular bisector l of the line AB where $A(1,3)$ and $B(5,9)$.

The line AB has gradient $\dfrac{9-3}{5-1} = \dfrac{3}{2}$

Hence l has gradient $-\dfrac{2}{3}$.

The midpoint $(3,6)$ of AB lies on l.

Hence l has equation $y - 6 = -\dfrac{2}{3}(x - 3)$

i.e. $\qquad 3y + 2x = 24$

3 Expand $(a + b)^n$ where n is a positive integer.

e.g. Expand $(a + b)^3$

The coefficients in the expansion are given by

Pascal's triangle:
```
        1
      1   1
    1   2   1
  1   3   3   1
```

Hence $(a + b)^3 = a^3 + 3a^2b + 3ab^2 + b^3$

Check in:

1 Express these complex numbers in exact modulus–argument form. Arguments should be principal and in radians.

a $z = 4 - 4i$

b $z = -1 + i\sqrt{3}$

c $z = -\sqrt{12} - 2i$

d $z = \dfrac{1}{3} + \dfrac{1}{\sqrt{3}}i$

2 a Find an equation for the perpendicular bisector of the line AB when

   i $A(2,7)$ and $B(2,8)$

   ii $A(-2,5)$ and $B(2,8)$

   iii $A\left(\dfrac{3}{2}, \dfrac{7}{4}\right)$ and $B\left(\dfrac{1}{2}, -\dfrac{3}{4}\right)$.

b The points $P(9,4)$, $Q(6,5)$ and $R(2,3)$ lie on a common circle C. By finding equations for the perpendicular bisectors of PQ and QR, or otherwise, find the centre and radius of circle C.

3 Expand

a $(a + b)^4$

b $(p + 2q)^3$

c $(a - b)^5$

FP2

By definition, the complex number
$$e^{i\theta} = \cos\theta + i\sin\theta$$
where θ is measured in radians.

This relationship is in the **FP2** section of the formula book.

EXAMPLE 1

Evaluate $\left(e^{i\pi}\right)^2$

$$\left(e^{i\pi}\right)^2 = e^{i2\pi}$$

Apply the definition $e^{i\theta} \equiv \cos\theta + i\sin\theta$ when $\theta = 2\pi$:

$$e^{i2\pi} = \cos(2\pi) + i\sin(2\pi)$$
$$= 1 + 0i$$

So $\left(e^{i\pi}\right)^2 = 1$

The usual rules of indices apply.

You can express a complex number z with modulus r and argument θ in the form
$$z = r(\cos\theta + i\sin\theta)$$
Since $\cos\theta + i\sin\theta \equiv e^{i\theta}$, you can write $z = re^{i\theta}$

See **FP1** for revision of modulus–argument form.

The complex number z has exponential form $z = re^{i\theta}$, where r is the modulus of z and θ is an argument of z.

EXAMPLE 2

Express $z = 2 + 2i$ in exponential form.

Find the modulus and principal argument of z:

$$r = |z| = \sqrt{8} \qquad \arg z = \frac{1}{4}\pi$$

So, in exponential form, $z = \sqrt{8}e^{i\frac{1}{4}\pi}$

$\frac{1}{4}\pi$ is the principal argument of z.

$\sqrt{8}e^{i\left(\frac{1}{4}\pi + 2\pi\right)}$ is also an exponential form for z since adding 2π to $\arg z$ produces the same complex number.

If the complex number z has modulus r and argument θ then the general exponential form for z is $z = re^{i(\theta + 2\pi k)}$, $k \in \mathbb{Z}$.

EXAMPLE 3

Find a general expression in exponential form for the complex number $1 + i\sqrt{3}$

$z = 1 + i\sqrt{3}$ has modulus $r = |z| = 2$

and principal argument $\arg z = \frac{1}{3}\pi$

Hence $\left(\frac{1}{3}\pi + 2\pi k\right)$, where k is any integer, is also an argument of z.

The general exponential form for z is $2e^{\left(\frac{1}{3}\pi + 2\pi k\right)i}$, $k \in \mathbb{Z}$.

Adding a multiple of 2π to $\arg z$ leaves z unmoved on the Argand diagram.

You can use exponential forms to deduce rules for the moduli and arguments of complex numbers.

EXAMPLE 4

If z, w are two non-zero complex numbers, show that

a $|zw| = |z||w|$

b $\arg(zw) = \arg z + \arg w$

Express z and w in exponential form:

$$z = re^{i\theta}, \quad w = se^{i\varphi}$$

Then $zw = \left(re^{i\theta}\right)\left(se^{i\varphi}\right)$

Simplify, adding the indices as usual:

$$= rse^{i(\theta + \varphi)}$$

So in exponential form, $zw = rse^{i(\theta + \varphi)}$

$|z| = r$, $\arg z = \theta$
$|w| = s$, $\arg w = \varphi$

a The modulus of zw is rs, where $r = |z|$ and $s = |w|$,
i.e. $|zw| = |z||w|$

b An argument of zw is $\theta + \varphi$, where $\theta = \arg z$ and $\varphi = \arg w$,
$\arg(zw) = \arg z + \arg w$

If z and w are any non-zero complex numbers then

$$|zw| = |z||w| \qquad \left|\frac{z}{w}\right| = \frac{|z|}{|w|}$$

$\arg(zw) = \arg z + \arg w$ ($\pm 2\pi$ to give principal value if necessary)

$\arg\left(\dfrac{z}{w}\right) = \arg z - \arg w$ ($\pm 2\pi$ to give principal value if necessary)

$|zw| = |z||w|$ is also true when either $z = 0$ or $w = 0$ (or both).

Arguments behave like logarithms.

Try to prove these results using the techniques from Example 4.

You can apply these rules to simplify products and quotients of complex numbers.

FP2

EXAMPLE 5

Express $\dfrac{8\left(\cos\frac{5}{9}\pi + i\sin\frac{5}{9}\pi\right)}{2\left(\cos\frac{2}{9}\pi + i\sin\frac{2}{9}\pi\right)}$ in the form $a + bi$ for exact $a, b \in \mathbb{R}$.

Identify the modulus and argument of each complex number in the product:

$z = 8\left(\cos\frac{5}{9}\pi + i\sin\frac{5}{9}\pi\right)$ has modulus 8 and argument $\frac{5}{9}\pi$

$w = 2\left(\cos\frac{2}{9}\pi + i\sin\frac{2}{9}\pi\right)$ has modulus 2 and argument $\frac{2}{9}\pi$

So $\left|\dfrac{z}{w}\right| = \dfrac{|z|}{|w|}$ and $\arg\left(\dfrac{z}{w}\right) = \arg z - \arg w$

$\qquad = \dfrac{8}{2} = 4 \qquad\qquad\quad = \dfrac{5}{9}\pi - \dfrac{2}{9}\pi = \dfrac{1}{3}\pi$

Hence the quotient $\dfrac{z}{w}$ has modulus 4 and argument $\frac{1}{3}\pi$

i.e. $\dfrac{8\left(\cos\frac{5}{9}\pi + i\sin\frac{5}{9}\pi\right)}{2\left(\cos\frac{2}{9}\pi + i\sin\frac{2}{9}\pi\right)} = 4\left(\cos\frac{1}{3}\pi + i\sin\frac{1}{3}\pi\right) = 2 + i2\sqrt{3}$ $\cos\frac{1}{3}\pi = \frac{1}{2}, \sin\frac{1}{3}\pi = \frac{\sqrt{3}}{2}$

Exercise 3.1

1 Express each of these numbers in the form $x + iy$ for exact $x, y \in \mathbb{R}$.

a $e^{\frac{1}{2}\pi i}$

b $e^{3\pi i}$

c $e^{\frac{1}{6}\pi i}$

d $e^{-\frac{\pi}{4}i}$

e $e^{\frac{1}{\frac{1}{3}\pi i}}$

f $\left(e^{\frac{1}{3}\pi i}\right)^{2}$

2 Express these numbers in modulus–argument form.

a $3e^{\frac{2}{7}\pi i}$

b $\dfrac{1}{2e^{\frac{1}{5}\pi i}}$

c $4e^{2i}$

3 Express these numbers in the form $re^{i\theta}$ for exact $r > 0$ and $-\pi < \theta \leqslant \pi$.

a $1 + i\sqrt{3}$

b $2\sqrt{3} - 2i$

c $-4 + 4i$

d $-\sqrt{6} - i\sqrt{2}$

e 5

f $-8i$

4 Express these in the form $x + iy$ for exact $x, y \in \mathbb{R}$.

a $\left(\cos\frac{2}{5}\pi + i\sin\frac{2}{5}\pi\right)\left(\cos\frac{3}{5}\pi + i\sin\frac{3}{5}\pi\right)$

b $\dfrac{\left(\cos\frac{5}{6}\pi + i\sin\frac{5}{6}\pi\right)}{\left(\cos\frac{2}{3}\pi + i\sin\frac{2}{3}\pi\right)}$

c $2\left(\cos\frac{3}{10}\pi + i\sin\frac{3}{10}\pi\right) \times 3\left(\cos\frac{1}{5}\pi + i\sin\frac{1}{5}\pi\right)$

d $\dfrac{3\left(\cos\frac{2}{9}\pi + i\sin\frac{2}{9}\pi\right)}{4\left(\cos\frac{1}{18}\pi + i\sin\frac{1}{18}\pi\right)}$

5 Simplify these expressions.

   a $(\cos 2\theta + i \sin 2\theta)(\cos 3\theta + i \sin 3\theta)$

   b $\dfrac{4(\cos 6\theta + i \sin 6\theta)}{2(\cos 2\theta + i \sin 2\theta)}$

   c $(\cos \theta + i \sin \theta)^2$

   d $(2(\cos \theta + i \sin \theta))^3$

6 z and w are complex numbers, where $z = e^{i\frac{2}{3}\pi}$, such that zw is a real negative number.

   a Write down $\arg z$.

   b Hence find, in terms of π, the principal argument of w.

7 It is given that $\dfrac{z}{w}$ is a purely imaginary number, and $z = re^{\frac{1}{4}\pi i}$, where $r > 0$.
Find, in terms of π, the two possible principal arguments of w.

8 Given that the complex number $z = re^{i\theta}$, express the following in exponential form. Give answers in terms of r and θ.

   a z^2 **b** z^* **c** $\dfrac{1}{z}$

> z^* is the complex conjugate of z. If $z = a + bi$ then $z^* = a - bi$. See **FP1**.

9 Express these numbers in general exponential form.

   a $3 + 3i$ **b** $\sqrt{15} - i\sqrt{5}$ **c** $-2 + i\sqrt{12}$

10 Prove that

   a $(\cos \theta + i \sin \theta)(\cos \varphi + i \sin \varphi) = (\cos(\theta + \varphi) + i \sin(\theta + \varphi))$

   b $\dfrac{\cos \theta + i \sin \theta}{\cos \varphi + i \sin \varphi} - \cos(\theta - \varphi) + i \sin(\theta - \varphi)$

11 Given that $z = e^{i\theta}$, where θ is acute, find in exponential form

   a iz **b** $z + iz$ **c** $1 + z^2$ **d** $1 - z^2$
   Give each answer in terms of θ.

> For **c** use a trigonometric identity.

12 In this question you may assume the usual rules of indices apply throughout.

   a Evaluate $e^{i\left(\frac{1}{2}\pi + 2\pi k\right)}$ where k is any integer.

   b Hence show that $i^i = e^{-\frac{\pi}{2}(4k+1)}$ where k is any integer.

   c Deduce the values of $(-1)^i$.

13 In this question you may treat the imaginary number i as a constant.
Let $y = \cos \theta + i \sin \theta$.

   a State the value of y when $\theta = 0$.

   b Show that $\dfrac{dy}{d\theta} = iy$

   c Solve this differential equation, expressing y in terms of θ.

   d Comment on whether the argument proves that $e^{i\theta} \equiv \cos \theta + i \sin \theta$

FP2

3.2 Solving equations using exponential forms

You can solve an equation of the form $z^n = \alpha$, where n is a positive integer and α is any real or complex number. The solution consists of n numbers, each called an *nth root* of α.

$z^n = \alpha$ so $z = \sqrt[n]{\alpha}$

EXAMPLE 1

a Solve the equation $z^3 = 8$,
 giving answers in exact cartesian form.

b Hence display the three cube roots of 8 on an
 Argand diagram.

---

a Express z in exponential form $z = re^{i\theta}$, where $r = |z| \in \mathbb{R}$, $\theta = \arg z$:

The arguments of the real number 8 are $0, \pm 2\pi, \pm 4\pi, \ldots$
so the most general exponential form for 8 is $8e^{2\pi ki}$,
where $k \in \mathbb{R}$.

> You need to use the general form to find all the cube roots of 8.

Express the equation $z^3 = 8$ using exponential forms:

$$\left(re^{i\theta}\right)^3 = 8e^{2\pi ki}$$

i.e. $\quad r^3 e^{3\theta i} = 8e^{2\pi ki}$

Compare the moduli: $\quad r^3 = 8 \quad$ so $\quad r = 2$

r is the real cube root of 8.

Compare arguments: $\quad 3\theta = 2\pi k \quad$ so $\quad \theta = \dfrac{2\pi k}{3}$

So the solution to the equation $z^3 = 8$ is given by

$$z = 2e^{\left(\frac{2\pi k}{3}\right)i}, k \in \mathbb{Z}$$

$z = re^{i\theta}$ so $z = 2e^{\left(\frac{2\pi k}{3}\right)i}, k \in \mathbb{Z}$

Find each value of z by letting k take integer values, starting with $k = 0$:

When $\quad k = 0$: $z_0 = 2e^{\left(\frac{2\pi \times 0}{3}\right)i} = 2(\cos 0 + i\sin 0)$
$$= 2 + 0i$$

This is the real cube root of 8.

$$k = 1: z_1 = 2e^{\frac{2}{3}\pi i} = 2\left(\cos\frac{2}{3}\pi + i\sin\frac{2}{3}\pi\right)$$

$$= -1 + i\sqrt{3}$$

$$k = 2: z_2 = 2e^{\frac{4}{3}\pi i} = 2\left(\cos\frac{4}{3}\pi + i\sin\frac{4}{3}\pi\right)$$

$$= -1 - i\sqrt{3}$$

z_1 and z_2 are the complex cube roots of 8.

There can only be three distinct values of z since
$z^3 = 8$ is a cubic equation.
So the solution of $z^3 = 8$ is given by
$$z_0 = 2, \quad z_1 = -1 + i\sqrt{3}, \quad z_2 = -1 - i\sqrt{3}$$

$k = 3$ gives the same root as $k = 0$:

$$2e^{\left(\frac{2\pi \times 3}{3}\right)i} = 2e^{2\pi i} = 2e^{0i} = z_0$$

EXAMPLE 1 (CONT.)

b The values z_0, z_1 and z_2 in part **a** each satisfy the equation $z^3 = 8$ and are therefore the three cube roots of 8.

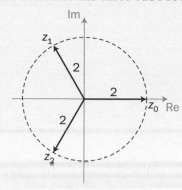

The roots are equally spaced around the unit circle, separated by angles of $\frac{2\pi}{3}$ radians.

EXAMPLE 2

Find the fourth roots of 4i.

Give each answer in the form $re^{i\theta}$ where $r > 0$ and $-\pi < \theta \leqslant \pi$ are exact.

Solve the equation $z^4 = 4i$ to find the fourth roots of 4i using exponential forms:

Use the general exponential form of $4i = 4e^{\left(\frac{1}{2}\pi + 2\pi k\right)i}$, where $k \in \mathbb{Z}$.

Since $z^4 = 4i$

$$r^4 e^{4\theta i} = 4e^{\left(\frac{1}{2}\pi + 2\pi k\right)i}$$

$|4i| = 4$, $\arg(4i) = \frac{1}{2}\pi$

Compare the moduli: $r^4 = 4$ so $r = \sqrt{2}$

$r > 0$ so r is the *positive* 4th root of 4.

Compare arguments: $4\theta = \frac{1}{2}\pi + 2\pi k$ so $\theta = \left(\frac{1+4k}{8}\right)\pi$

Hence $z = \sqrt{2}e^{\left(\frac{1+4k}{8}\right)\pi i}$, $k \in \mathbb{Z}$

Find each value of z by letting k take integer values starting with $k = 0$:

When $k = 0$: $z_0 = \sqrt{2}e^{\frac{1}{8}\pi i}$

When $k = 1$: $z_1 = \sqrt{2}e^{\frac{5}{8}\pi i}$

When $k = 2$, $\theta = \frac{9}{8}\pi > \pi$

Use *negative* integers for k to obtain suitable values of θ:

When $k = -1$: $z_2 = \sqrt{2}e^{-\frac{3}{8}\pi i}$

$-\pi < -\frac{3}{8}\pi \leqslant \pi$

When $k = -2$: $z_3 = \sqrt{2}e^{-\frac{7}{8}\pi i}$

$-\pi < -\frac{7}{8}\pi \leqslant \pi$

Hence the four fourth roots of 4i are

$$z_0 = \sqrt{2}e^{\frac{1}{8}\pi i}, \quad z_1 = \sqrt{2}e^{\frac{5}{8}\pi i}, \quad z_2 = \sqrt{2}e^{-\frac{3}{8}\pi i}, \quad z_3 = \sqrt{2}e^{-\frac{7}{8}\pi i}$$

FP2

You can show the fourth roots of 4i, as found in Example 2, on an Argand diagram:

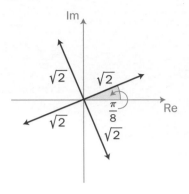

The roots are equally spaced around the unit circle, separated by angles of $\frac{2\pi}{4}$ radians.

You can evaluate expressions involving the roots of the number 1, known as the **roots of unity**.

It is given that w is a complex cube root of 1.

a Show that 1, w and w^2 are the three cube roots of 1.

b Evaluate $1 + w + w^2$

---

a As there are only three cube roots of 1, any three distinct numbers each of whose cube is 1 must form a complete list of the cube roots of unity.

$1^3 = 1$ and, by definition, $w^3 = 1$

Evaluate $(w^2)^3$ to show w^2 is also a cube root of 1:

$$\left(w^2\right)^3 = \left(w^3\right)^2$$
$$= (1)^2$$
$$= 1$$

Hence the three numbers 1, w and w^2 are the cube roots of unity.

Rules of indices:
$(a^n)^m = a^{nm} = (a^m)^n$

The numbers 1, w and w^2 are all different.

b $1 + w + w^2 = \dfrac{w^3 - 1}{w - 1}$

$$= \dfrac{0}{w - 1} = 0$$

$1 + w + w^2$ is a geometric series, with first term 1 and common ratio w.

$w^3 = 1$ so $w^3 - 1 = 0$

The equation $z^n = 1$ has solution $z = e^{\frac{2\pi k}{n}i}$, $0 \leqslant k \leqslant n-1$.
The values of z are the nth roots of unity.
If w is any complex nth root of unity then $1, w, w^2, \ldots, w^{n-1}$ are all the nth roots of unity.
The sum of these roots is zero (i.e. $1 + w + w^2 + \cdots + w^{n-1} = 0$).

The nth roots of unity are given in the **FP2** section of the formula book.

FP2

Exercise 3.2

1 Solve each of these equations.
Give answers in the form $re^{i\theta}$ where $r > 0$ and $0 \leqslant \theta < 2\pi$ are exact.

a $z^3 = 27$

b $z^4 = -1$

c $z^5 = 32i$

d $8z^6 = 1$

2 Solve each of these equations.
Give answers in the form $re^{i\theta}$ where $r > 0$ and $-\pi \leqslant \theta < \pi$ are exact.

a $z^3 = 2 + 2i$

b $z^4 = 8 + 8i\sqrt{3}$

c $z^3 - \dfrac{1}{4} - \dfrac{1}{4}i$

d $z^7 - \dfrac{2i}{1-i}$

3 a Solve the equation $z^3 = \dfrac{1}{8}$
Give answers in exact cartesian form.

b Illustrate the three cube roots of $\dfrac{1}{8}$ on a single Argand diagram.

c Give a geometrical reason why the sum of these three roots is zero.

4 a Find the fourth roots of 9.
Give answers in cartesian form.

b Illustrate these roots on a single Argand diagram.

c Show that the points representing these roots are the vertices of a square and find the area of this square.

5 a Solve the equation $z^6 = 64$
Give answers in the form $re^{i\theta}$ where $r > 0$ and $-\pi \leqslant \theta < \pi$ are exact.

b Illustrate on a single Argand diagram the 6th roots of 64.

The number w is the sixth root of 64 with smallest positive argument.

c Using your diagram, or otherwise, show that $|w - 2| = 2$

6 a Solve the equation $z^3 = \sqrt{2} + i\sqrt{6}$
Give answers in modulus–argument form.

b Illustrate the cube roots of $\sqrt{2} + i\sqrt{6}$ on a single Argand diagram.

c Find the exact area of the triangle whose vertices are the points on the Argand diagram which represent these cube roots.

7 Given that w is a complex cube root of unity, evaluate

  a $w(1 + w)$ **b** $(w^2 + 1)(1 + w)$

  c $(w + 1)^3$ **d** $\dfrac{1}{1+w} + w$

8 Given that w is a complex fifth root of 1, evaluate

  a $(1 + w)(1 - w + wv^2 - w^3 + w^4)$

  b $w(w + 1)(w^2 + 1)$

  c $\left(\dfrac{1}{w} + 1\right)\left(\dfrac{1}{w} + w^2\right)$

9 **a**   Solve the equation $z^4 = \dfrac{1}{2} + i\dfrac{\sqrt{3}}{2}$. Give answers in the form

      $re^{i\theta}$ where $r > 0$ and $-\pi < \theta \leqslant \pi$ are exact.

  b   Illustrate these roots on an Argand diagram.

  c   Find the two possible values of $|p + q|$ where p and q are any
      pair of distinct roots of this equation.

10   In the Argand diagram, point A represents the complex number $a = -\dfrac{1}{8} + \dfrac{1}{8}i$

  a   Solve the equation $z^5 = a$. Give answers in the form $re^{i\theta}$
     where $r > 0$ and $-\pi < \theta \leqslant \pi$ is exact.

  b   On a single Argand diagram show point A and the roots of the equation $z^5 = a$

  c   Express the root of this equation which lies in the fourth quadrant
     in the form λa, stating the value of the real constant λ.

11   The vertices A, B, C, D and E of the regular pentagon shown
   on the Argand diagram represent the roots of the equation $z^n = 1$,
   where n is a positive integer.

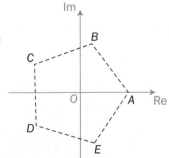

  a   State the value of n.

  b   Hence write down, in modulus–argument form,
     the complex number represented by the point B.

  c   Show that the exact perimeter of the pentagon
     $ABCDE$ is $10\sin\left(\dfrac{1}{5}\pi\right)$.

12 a Find the two roots of the equation $z^6 + 8i = 0$ which lie in the second quadrant of the Argand diagram. Give answers in the form $re^{i\theta}$ where $r > 0$ and $-\pi < \theta \leqslant \pi$ is exact.

On the Argand diagram, points A and B represent these two roots. Point M is the midpoint of the line AB.

b Find the complex number which represents point M. Give your answer in the form $a + bi$ where $a, b \in \mathbb{R}$ are in simplified surd form.

13 For a positive integer n, it is given that z and w are two distinct complex nth roots of unity.

a Use an algebraic approach to show that zw is also a nth root of unity.

b Use a geometrical approach to show that $|z + w| < 2$

c Given that $|z + w| = \sqrt{2}$, show that n is a multiple of 4.

14 The Argand diagram shows w, the complex nth root of unity with smallest positive argument, where $n \in \mathbb{Z}$, $n > 4$

a Write down arg w, giving your answer in terms of n and hence explain why w must lie in the first quadrant.

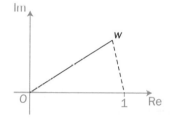

b Show that

i $|w - 1| = 2\sin\left(\dfrac{\pi}{n}\right)$

ii $\arg(w - 1) - \left(\dfrac{n+2}{2n}\right)\pi$

c Show that $w + 1 = 2\cos\left(\dfrac{\pi}{n}\right)e^{i\frac{\pi}{n}}$

Use a geometrical approach.

d Hence, or otherwise, show that

$\left|w^2 - 1\right| = 2\sin\left(\dfrac{2\pi}{n}\right)$

and find an expression for $\arg(w^2 - 1)$ in terms of n.

$e^{i\theta} \equiv \cos\theta + i\sin\theta$

Hence $(\cos\theta + i\sin\theta)^2 \equiv (e^{i\theta})^2$
$$\equiv e^{i(2\theta)}$$
$$\equiv \cos 2\theta + i\sin 2\theta$$

Rules of indices: $(e^a)^b = e^{ab}$

De Moivre's theorem states that, in general,
$$\{r(\cos\theta + i\sin\theta)\}^n \equiv r^n(\cos n\theta + i\sin n\theta)$$
where n is any integer.

This result is given in the **FP2** section of the formula book.

EXAMPLE 1

Use de Moivre's theorem to express these numbers in cartesian form.

a $\left(\cos\frac{1}{9}\pi + i\sin\frac{1}{9}\pi\right)^3$

b $\left(2\left(\cos\frac{3}{8}\pi + i\sin\frac{3}{8}\pi\right)\right)^{-4}$

---

a $(\cos\theta + i\sin\theta)^n \equiv \cos n\theta + i\sin n\theta$

So $\left(\cos\frac{1}{9}\pi + i\sin\frac{1}{9}\pi\right)^3 = \cos\left(3 \times \frac{1}{9}\pi\right) + i\sin\left(3 \times \frac{1}{9}\pi\right)$

$$= \cos\frac{1}{3}\pi + i\sin\frac{1}{3}\pi$$

$$= \frac{1}{2} + \frac{1}{2}\sqrt{3}i$$

b $\left(2\left(\cos\frac{3}{8}\pi + i\sin\frac{3}{8}\pi\right)\right)^{-4} = 2^{-4}\left(\cos\left(-\frac{3}{2}\pi\right) + i\sin\left(-\frac{3}{2}\pi\right)\right)$

$$= \frac{1}{16}(0 + 1i)$$

$$= \frac{1}{16}i$$

$\cos\left(-4 \times \frac{3}{8}\pi\right) = \cos\left(-\frac{3}{2}\pi\right) = 0$

$\sin\left(-4 \times \frac{3}{8}\pi\right) = \sin\left(-\frac{3}{2}\pi\right) = 1$

You can only apply de Moivre's theorem to an expression of the form $\{r(\cos\theta + i\sin\theta)\}^n$.

Use de Moivre's theorem to evaluate $\left(\cos\frac{2}{3}\pi - i\sin\frac{2}{3}\pi\right)^6$

Express $\cos\frac{2}{3}\pi - i\sin\frac{2}{3}\pi$ in the form $\cos\theta + i\sin\theta$:

$$\cos\left(-\frac{2}{3}\pi\right) = \cos\frac{2}{3}\pi$$

$$\sin\left(-\frac{2}{3}\pi\right) = -\sin\frac{2}{3}\pi$$

Hence $\cos\frac{2}{3}\pi - i\sin\frac{2}{3}\pi = \cos\left(-\frac{2}{3}\pi\right) + i\sin\left(-\frac{2}{3}\pi\right)$

So $\left(\cos\frac{2}{3}\pi - i\sin\frac{2}{3}\pi\right)^6 = \left[\cos\left(-\frac{2}{3}\pi\right) + i\sin\left(-\frac{2}{3}\pi\right)\right]^6$

Use de Moivre's theorem: $= \cos\left(6 \times -\frac{2}{3}\pi\right) + i\sin\left(6 \times -\frac{2}{3}\pi\right)$

$$= \cos(-4\pi) + i\sin(-4\pi)$$

$$= 1 + i0 = 1$$

Check that you understand these results by looking at the graphs of $\cos\theta$ and $\sin\theta$.

The principal argument of $\cos\frac{2}{3}\pi - i\sin\frac{2}{3}\pi$ is $-\frac{2}{3}\pi$.

You can use de Moivre's theorem to calculate powers of a complex number.

Use de Moivre's theorem to express $\left(\sqrt{3} + i\right)^5$ in exact cartesian form.

Express $z = \sqrt{3} + i$ in modulus-argument form:

$|z| = 2$, $\arg z = \frac{1}{6}\pi$ so $z = 2\left(\cos\frac{1}{6}\pi + i\sin\frac{1}{6}\pi\right)$

So $z^5 = \left[2\left(\cos\frac{1}{6}\pi + i\sin\frac{1}{6}\pi\right)\right]^5$

$$= 2^5\left(\cos\left(5 \times \frac{1}{6}\pi\right) + i\sin\left(5 \times \frac{1}{6}\pi\right)\right)$$

$$= 32\left(-\frac{\sqrt{3}}{2} + \frac{1}{2}i\right)$$

$$= -16\sqrt{3} + 16i$$

Remember to raise 2 to the power of 5.

FP2

Exercise 3.3

1 Use de Moivre's theorem to express these numbers in exact cartesian form.

a $\left(\cos\frac{1}{2}\pi + i\sin\frac{1}{2}\pi\right)^6$

b $\left(\cos\frac{1}{6}\pi + i\sin\frac{1}{6}\pi\right)^3$

c $\left(\cos\frac{1}{4}\pi + i\sin\frac{1}{4}\pi\right)^6$

d $\left(2\left(\cos\frac{1}{12}\pi + i\sin\frac{1}{12}\pi\right)\right)^4$

e $\left(\cos\frac{2}{3}\pi + i\sin\frac{2}{3}\pi\right)^{-2}$

f $\left(\sqrt{2}\left(\cos\frac{3}{4}\pi + i\sin\frac{3}{4}\pi\right)\right)^{-3}$

2 Use de Moivre's theorem to express these numbers in exact cartesian form.

a $\left(\cos\frac{1}{3}\pi - i\sin\frac{1}{3}\pi\right)^7$

b $\dfrac{1}{\left(\cos\frac{1}{6}\pi - i\sin\frac{1}{6}\pi\right)^4}$

3 Simplify each of these. Give answers in exact cartesian form where appropriate.

a $\left(\cos\frac{1}{12}\pi + i\sin\frac{1}{12}\pi\right)^3 \times \left(\cos\frac{1}{8}\pi + i\sin\frac{1}{8}\pi\right)^2$

b $\dfrac{8\left(\cos\frac{1}{2}\pi + i\sin\frac{1}{2}\pi\right)^3}{2\left(\cos\frac{1}{4}\pi + i\sin\frac{1}{4}\pi\right)^5}$

c $\sqrt{2}\left(\cos\frac{2}{3}\pi + i\sin\frac{2}{3}\pi\right)^7 \times \sqrt{2}\left(\cos\frac{1}{3}\pi - i\sin\frac{1}{3}\pi\right)^5$

d $\dfrac{\left(\cos\frac{3}{8}\pi + i\sin\frac{3}{8}\pi\right)^2}{\left(\cos\frac{1}{4}\pi - i\sin\frac{1}{4}\pi\right)^5}$

4 Use de Moivre's theorem to evaluate each of these numbers. Give answers in exact cartesian form where appropriate.

a $\left(1 + i\sqrt{3}\right)^{12}$

b $(2 - 2i)^6$

c $\left(-\sqrt{6} + i\sqrt{2}\right)^4$

d $\left(\frac{\sqrt{3}}{2} + \frac{1}{2}i\right)^{17}$

5 a Express $1 + i$ in modulus–argument form.

b Hence show that when $n = 4m$, for m an odd integer, $(1 + i)^n = -4^m$

6 $z = \lambda + \sqrt{3}\lambda i$ where $\lambda > 0$ is a constant.

a Show that $z^n = 2^n\lambda^n\left(\cos\frac{n\pi}{3} + i\sin\frac{n\pi}{3}\right)$ for any integer n.

b Given that $z^6 = 8$, find the exact value of λ.

c Find the value of z^{-3}.

7 Use de Moivre's theorem to

 a prove that

 $$(\cos\theta - i\sin\theta)^n \equiv \cos n\theta - i\sin n\theta$$

 for n any integer

 b verify that $\cos\dfrac{p}{q}\theta + i\sin\dfrac{p}{q}\theta$ is a possible value of

 $(\cos\theta + i\sin\theta)^{\frac{p}{q}}$, where p and q are integers, $q \neq 0$.

8 a Simplify $\cos\left(\dfrac{\pi}{2} - \theta\right)$

 b Hence show that $(\sin\theta + i\cos\theta)^6 = -\cos 6\theta + i\sin 6\theta$

9 a Use induction to prove that

 $$(\cos\theta + i\sin\theta)^n \equiv \cos n\theta + i\sin n\theta$$

 for all positive integers n.

In the exam, you may be asked to prove de Moivre's theorem by induction.

 b Deduce that

 $$(\cos\theta + i\sin\theta)^m \equiv \cos m\theta + i\sin m\theta$$

 holds for all negative integers m.

Write $m = -n$, where $n > 0$.

FP2

10 In the Argand diagram the point P_1 represents the complex number

 $$z = r(\cos\theta + i\sin\theta), \text{ where } r > 0 \text{ and } \tfrac{1}{4}\pi < \theta < \tfrac{1}{2}\pi$$

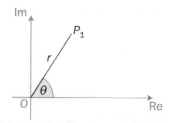

 a Write down z^2 in modulus–argument form, giving your answer in terms of r and θ.

Point P_2 represents the complex number z^2.

 b State the quadrant containing P_2.

 c Describe, in terms of an appropriate stretch and rotation, how the vector $\overrightarrow{OP_1}$ can be transformed into the vector $\overrightarrow{OP_2}$.

 d Show that $|z - 1|^2 = r^2 + 1 - 2r\cos\theta$

The point P_n represents the complex number z^n, where n is a positive integer.

 e On separate diagrams, sketch a curve which passes through the points P_n as n increases when
  i $r > 1$ ii $r < 1$ iii $r = 1$

$$e^{i\theta} \equiv \cos\theta + i\sin\theta \qquad \text{and} \qquad e^{-i\theta} \equiv \cos\theta - i\sin\theta$$

Add these results:　　　　　　　　　**Subtract these results:**

$$e^{i\theta} + e^{-i\theta} \equiv 2\cos\theta \quad (1) \qquad\qquad e^{i\theta} - e^{-i\theta} \equiv 2i\sin\theta \quad (2)$$

More generally, for any integer $n \geqslant 1$,

$$e^{in\theta} + e^{-in\theta} \equiv 2\cos n\theta \quad (3) \qquad\qquad e^{in\theta} - e^{-in\theta} \equiv 2i\sin n\theta \quad (4)$$

Using de Moivre's theorem.

You can use these results to prove trigonometric identities.

EXAMPLE 1

Prove that $4\cos^3\theta \equiv \cos 3\theta + 3\cos\theta$

Use (1): $2\cos\theta \equiv e^{i\theta} + e^{-i\theta}$:
So $(2\cos\theta)^3 \equiv (e^{i\theta} + e^{-i\theta})^3$
$$\equiv (e^{i\theta})^3 + 3(e^{i\theta})^2(e^{-i\theta}) + 3(e^{i\theta})(e^{-i\theta})^2 + (e^{-i\theta})^3$$

Simplify, using the rules of indices:
$$\equiv e^{i3\theta} + 3e^{i\theta} + 3e^{-i\theta} + e^{-i3\theta}$$

Group like terms:
$$\equiv (e^{i3\theta} + e^{-i3\theta}) + 3(e^{i\theta} + e^{-i\theta})$$
$$\equiv (2\cos 3\theta) + 3(2\cos\theta)$$

i.e. $(2\cos\theta)^3 \equiv 2\cos 3\theta + 6\cos\theta$

So $8\cos^3\theta = 2\cos 3\theta + 6\cos\theta$
i.e. $4\cos^3\theta = \cos 3\theta + 3\cos\theta$

Use (1) since the LHS in the question involves $\cos^3\theta$.

See **C2** for revision of binomial expansions
$(a + b)^3 \equiv a^3 + 3a^2b + 3ab^2 + b^3$

$e^{i3\theta} + e^{-i3\theta} \equiv 2\cos 3\theta$
using (3) when $n = 3$.

$(2\cos\theta)^3 = 8\cos^3\theta$

You can use de Moivre's theorem to prove identities for $\cos n\theta$ and $\sin n\theta$.

EXAMPLE 2

a Use de Moivre's theorem to prove that
$\sin 3\theta \equiv 3\sin \theta - 4\sin^3 \theta$

b Hence solve the equation $\sin 3\theta = \sin \theta$ for $0 \leqslant \theta \leqslant \pi$

---

a de Moivre's theorem states that
$$\cos 3\theta + i\sin 3\theta = (\cos \theta + i\sin \theta)^3$$

Expand $(\cos \theta + i\sin \theta)^3$ to find its real and imaginary parts:
$$(\cos \theta + i\sin \theta)^3$$
$$\equiv \cos^3 \theta + 3\cos^2 \theta (i\sin \theta) + 3\cos \theta (i\sin \theta)^2 + (i\sin \theta)^3$$
$$\equiv \cos^3 \theta + 3i\cos^2 \theta \sin \theta - 3\cos \theta \sin^2 \theta - i\sin^3 \theta$$
$$\equiv \cos^3 \theta - 3\cos \theta \sin^2 \theta + i(3\cos^2 \theta \sin \theta - \sin^3 \theta)$$

See **C2** for revision of binomial expansions.

$\cos 3\theta + i\sin 3\theta$
$$\equiv (\cos^3 \theta - 3\cos \theta \sin^2 \theta) + i(3\cos^2 \theta \sin \theta - \sin^3 \theta)$$

Compare imaginary parts: $\sin 3\theta = 3\cos^2 \theta \sin \theta - \sin^3 \theta$

Use $\cos^2 \theta \equiv 1 - \sin^2 \theta$ to obtain an expression for $\sin 3\theta$ in terms of powers of $\sin \theta$ only:

$$\sin 3\theta \equiv 3\cos^2 \theta \sin \theta - \sin^3 \theta$$
$$\equiv 3(1 - \sin^2 \theta)\sin \theta - \sin^3 \theta$$

i.e. $\sin 3\theta \equiv 3\sin \theta - 4\sin^3 \theta$

b In the equation $\sin 3\theta = \sin \theta$, replace $\sin 3\theta$ with $3\sin \theta - 4\sin^3 \theta$ from a:

so $3\sin \theta - 4\sin^3 \theta = \sin \theta$

i.e. $2\sin \theta (1 - 2\sin^2 \theta) = 0$

Hence $\sin \theta = 0$ or $1 - 2\sin^2 \theta = 0$

The equation $\sin \theta = 0$ has solution $\theta = 0, \pi$.

The equation $1 - 2\sin^2 \theta = 0$ has solution $\theta = \frac{1}{4}\pi, \frac{3}{4}\pi$

For $0 \leqslant \theta \leqslant \pi$. $\sin \theta = \frac{1}{\sqrt{2}}$

Hence the solution of the equation $\sin 3\theta = \sin \theta$
is $\theta = 0, \frac{1}{4}\pi, \frac{3}{4}\pi, \pi$.

FP2

Exercise 3.4

1 Starting with the identity $2\cos \theta \equiv e^{i\theta} + e^{-i\theta}$, and squaring both sides, prove that
$$\cos^2 \theta \equiv \frac{1}{2}(\cos 2\theta + 1)$$

2 **a** Starting with the identity $2i\sin \theta \equiv e^{i\theta} - e^{-i\theta}$, prove that
$$\sin^3 \theta \equiv \frac{1}{4}(3\sin \theta - \sin 3\theta)$$

b Hence, or otherwise, solve the equation
$$6\sin \theta - 2\sin 3\theta = 1 \quad \text{for } 0 \leqslant \theta \leqslant \pi$$

3 a Show that $8\cos^4\theta \equiv \cos 4\theta + 4\cos 2\theta + 3$

   b Hence, or otherwise, solve the equation
   $\cos 4\theta + 4\cos 2\theta = 5$ for $0 \leqslant \theta \leqslant 2\pi$

4 a Express $\sin^4\theta$ in the form $A\cos 4\theta + B\cos 2\theta + C$
   for constants A, B and C to be determined.

   b Hence, or otherwise, find $\int \sin^4\theta \, d\theta$

5 a Show that $32\cos^5\theta \equiv 2\cos 5\theta + 10\cos 3\theta + 20\cos\theta$

   b Evaluate $\displaystyle\int_0^{\frac{1}{6}\pi} 16\cos^5\theta \, d\theta$
   Give your answer in exact form.

6 a Express $16\sin^5\theta$ as the sines of multiples of θ.

   b Hence solve the equation $\sin 5\theta - 5\sin 3\theta + \sin\theta = 0$ for $0 \leqslant \theta \leqslant 2\pi$.

7 It is given that $z = \cos\theta + i\sin\theta$

   a Show that $z^n + \dfrac{1}{z^n} = 2\cos n\theta$, for any positive integer n.

   You may be tested on this approach in the exam.

   b Hence, by expanding $\left(z + \dfrac{1}{z}\right)^3$, express $8\cos^3\theta$ as cosines of multiples of θ.

   c Find $\displaystyle\int \sqrt{(2\cos 2\theta + 2)^3}\, d\theta$

8 Use de Moivre's theorem to prove that
   $\cos 2\theta \equiv \cos^2\theta - \sin^2\theta$

   See Example 2.

9 a Use de Moivre's theorem to show that
   $\cos 3\theta \equiv 4\cos^3\theta - 3\cos\theta$

   b Hence, or otherwise, solve the equation
   $\cos 3\theta + 2\cos\theta = 0$ for $0 \leqslant \theta \leqslant \pi$.

   c Express $4\sin 3\theta \cos\theta(4\cos^2\theta - 3)$ in the form $A\sin B\theta$ for
   integers A and B to be stated.

10 **a** Use de Moivre's theorem to prove that
$\sin 3\theta \equiv 3\cos^2\theta - \sin^3\theta$

See Example 2 for help.

It is given that $\cos 3\theta \equiv \cos^3\theta - 3\cos\theta\sin^2\theta$

b Assuming $\cos\theta \neq 0$, show that

$$\tan 3\theta \equiv \frac{3\tan\theta - \tan^3\theta}{1 - 3\tan^2\theta}$$

11 **a** Use de Moivre's theorem to prove that

   i $\sin 4\theta \equiv 4\cos^3\theta\sin\theta - 4\cos\theta\sin^3\theta$

   ii $\cos 4\theta \equiv \cos^4\theta - 6\cos^2\theta\sin^2\theta + \sin^4\theta$

b Deduce that $\tan 4\theta = \dfrac{4\tan\theta - 4\tan^3\theta}{1 - 6\tan^2\theta + \tan^4\theta}$, for $\cos\theta \neq 0$

c Given that $\tan\theta = \sqrt{2}$ find the exact value of $\tan 4\theta$.
Give your answer in the form $q\sqrt{2}$ for q a rational
number to stated.

12 **a** Show that $\cos 5\theta \equiv \cos^5\theta - 10\cos^3\theta\sin^2\theta + 5\cos\theta\sin^4\theta$

b Hence express $\dfrac{\cos 5\theta}{\cos\theta}$ in terms of $\sin\theta$.

It is given that the smallest positive root k of the equation
$16x^4 - 12x^2 + 1 = 0$ lies in the interval $(0, 1)$.

c If $\theta = \arcsin k$ show that $k = \sin\left(\dfrac{1}{10}\pi\right)$

d Deduce that $\sin^2\left(\dfrac{1}{10}\pi\right) = \dfrac{3 - \sqrt{5}}{8}$

13 **a** Express $64\sin^6\theta$ as the cosines of multiples of θ.

The diagram shows part of the curve with equation $y = 8\sin^3\theta$

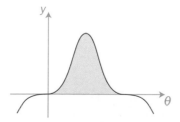

V is the volume formed when the region bounded by the curve and the
θ-axis, as shown on the diagram, is rotated once about the θ-axis.

b Show that $V = 20\pi^2$

c Express $64\sin^6\theta\cos^6\theta$ as cosines of multiples of θ.

FP2

3.5 Loci in the complex plane

The path of a point P which is governed by a rule is the **locus** of P. If P represents a complex number z then you can sketch the locus of P on an Argand diagram.

EXAMPLE 1

The point P represents a complex number z such that
$|z - 3| = 2$
Sketch the locus of P.

In the Argand diagram, $|z - 3|$ is the length of the line which joins P to the point representing the number 3.

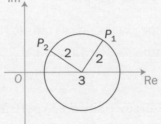

Refer to **FP1**.

So the locus $|z - 3| = 2$ describes all complex numbers z which lie 2 units from the fixed number 3.

The locus of P consists of precisely those points on the circle shown.

The locus of P is a circle with radius 2 and with centre $3 + 0i$

The cartesian equation of this locus is $(x - 3)^2 + y^2 = 4$

If P represents the complex number z for which $|z - a| = b$, where a is a fixed complex number and $b \in \mathbb{R}$, then the locus of P is a circle, centre a, radius b.

EXAMPLE 2

Sketch the locus of P, given that P represents the complex number z, where $|z - 4| = |z - 4i|$
Find the cartesian equation of this locus.

Represent the complex numbers 4 and 4 i by the points
A and B respectively:
$|z - 4| = AP$, $|z - 4i| = BP$

Since $|z - 4| = |z - 4i|$, $AP = BP$

The locus of P is the perpendicular bisector of the line AB.

Using the cartesian coordinates $A(4, 0)$ and $B(0, 4)$, the line AB has gradient -1 and midpoint $(2, 2)$.

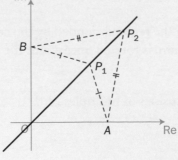

The locus of P consists of all points equidistant from A and B.

Hence the equation of the perpendicular bisector of AB is $y = x$

$\text{grad}_{AB} = \dfrac{4 - 0}{0 - 4} = -1$

If P represents the complex number z for which $|z - a| = |z - b|$, where a and b are fixed complex numbers, then the locus of P is the perpendicular bisector of the line AB, where points A and B represent a and b respectively.

You can sketch the locus $\arg(z - a) = \theta$, where a is a fixed complex number.

In the diagram, points P and A represent complex numbers z and a respectively. A line representing the complex number $z - a$ has been drawn.

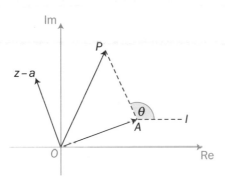

Refer to **FP1** Section 2.5. The vector $\overrightarrow{AP}$ is parallel to the line representing $z - a$

If $\arg(z - a) = \theta$, then P must be positioned so that θ is the angle between the vector $\overrightarrow{AP}$ and a line l parallel to the positive real axis.

FP2

EXAMPLE 3

Point P represents the complex number z such that
$$\arg(z - 2 - i) = \tfrac{1}{3}\pi$$
Sketch the locus of P.

Express $z - 2 - i$ in the form $z - a$:
$$z - 2 - i = z - (2 + i)$$

$a = 2 + i$

Represent the fixed complex number $2 + i$ by the point A and draw a line l from A parallel to the positive real axis:

The locus consists of points P for which the vector $\overrightarrow{AP}$ makes an angle of $\tfrac{1}{3}\pi$ against l.

The locus does not include the point A since the argument of the zero complex number is undefined.

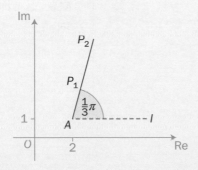

Do not extend the locus below the horizontal as points on this line are such that $\arg(z - 2 - i) = -\tfrac{2}{3}\pi$

If P represents the complex number z for which $\arg(z-a) = \theta$, where the complex number a is represented by the point A, then the locus of P is a half-line, beginning at A, inclined at an angle θ against the positive real axis.

You can describe a region on an Argand diagram using inequalities.

EXAMPLE 4

Shade the region on the Argand diagram given by

a $|z - 3 - i| < 3$

b $-\frac{1}{4}\pi \leqslant \arg z \leqslant \frac{1}{4}\pi$

-----

a Express $|z - 3 - i|$ in the form $|z - a|$:

$|z - 3 - i| = |z - (3 + i)|$

The inequality $|z - (3 + i)| < 3$ describes all complex numbers z which are *less* than 3 units from the fixed complex number $3 + i$. The region is the interior of a circle, centre $(3, 1)$, radius 3, as shaded in the diagram.

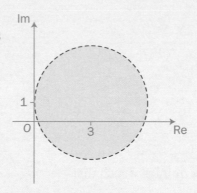

The circle is dotted to indicate that points on its circumference do not belong to the shaded region.

b $-\frac{1}{4}\pi \leqslant \arg z \leqslant \frac{1}{4}\pi$ describes the region bounded by two half-lines from O making an angle of $\frac{1}{4}\pi$ and $-\frac{1}{4}\pi$ respectively, against the positive real axis.

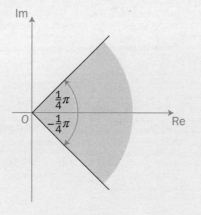

All points in the shaded region satisfy $-\frac{1}{4}\pi \leqslant \arg z \leqslant \frac{1}{4}\pi$

Exercise 3.5

1 The point P represents the complex number z. For each question, sketch the locus of P on an Argand diagram and find its cartesian equation.

   a $|z - 4| = 3$ **b** $|z + 6i| = 6$ **c** $|z - 3 + i| = 1$

   d $|z - 2 - 4i| = 3$ **e** $|z + 4 - i| = 2$ **f** $|z + 3 + 2i| = 4$

2 The point P represents the complex number z. For each question, sketch the locus of P on an Argand diagram and find its cartesian equation.

   a $|z + 4| = |z - 2i|$ **b** $|z - i| = |z + 3i|$ **c** $|z - 2 + i| = |z - 3 - 2i|$

   d $|z + 2| = |z + 5|$ **e** $|z + 3 + 3i| = |z + 2i|$ **f** $|z - 5 + 2i| = |z + 2 - 3i|$

3 Sketch, on separate Argand diagrams, the locus of the point P which represents the complex number z.

   a $\arg z = \frac{1}{6}\pi$ **b** $\arg z = \frac{2}{3}\pi$ **c** $\arg(z - 3) = \frac{1}{3}\pi$

   d $\arg(z - 2i) = -\frac{1}{4}\pi$ **e** $\arg(z - 3 + 2i) = \frac{1}{2}\pi$ **f** $\arg(z + 4 - 4i) = 1.5^c$

4 **a** Sketch these loci on the same Argand diagram.
   i $|2z - 3i| = 4$ **ii** $|2z - 3| = |2z - 5|$ Divide all terms by 2.

   b Write down the complex number which lies in both loci.

5 Describe the loci in these diagrams in terms of z.

a

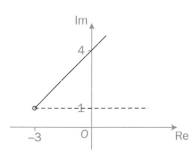

b

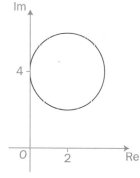

c

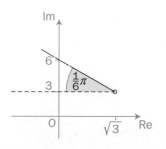

d

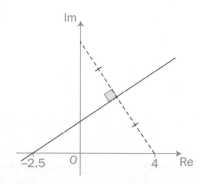

6 On separate Argand diagrams, shade the regions defined by each of these.

    a $|z - 4 + i| \leqslant 3$ **b** $|z + 2i| > 4$

    c $0 \leqslant \arg(z - 4 - 2i) \leqslant \frac{1}{2}\pi$ **d** $-\pi < \arg(z + 2 - i) \leqslant \frac{1}{4}\pi$

    e $|z - 3| \leqslant |z - 3i|$ **f** $|z - 2 + i| > |z - 4 + 3i|$

7 P represents the complex number z such that $|z - 4 - 4i| = 4$

    a Sketch the locus of P.

    b Find, in exact form, the greatest possible value of $|z|$, where z is any complex number in this locus.

8 On separate Argand diagrams, sketch the locus of points defined by

    a $\arg z = \tan^{-1}(2)$

    b $\tan(\arg (z - 1)) = \sqrt{3}$

    c $\tan^2(\arg (z + 2 - 3i)) = 1$

9 Shade, on the same diagram, the region satisfied by these loci.

    a $|z + 2 - 2i| \leqslant 2$ and $0 \leqslant \arg z \leqslant \frac{3}{4}\pi$

    b $|z - 3| \geqslant |z - 3i|$ and $0 \leqslant \arg(z - 2) < \frac{1}{2}\pi$

    c $|z + 4 + 4i| \leqslant 4$ and $|z + 2i| < |z + 4i|$

    d $|z + 3 - 3i| \leqslant |z|$ and $\frac{3}{4}\pi \leqslant \arg z \leqslant \frac{1}{2}\pi$

10 P represents the complex number z such that $|z| = 4$. Point Q represents the complex number z for which $\arg z = \frac{1}{4}\pi$.

    a Sketch, on the same Argand diagram, the locus of P and the locus of Q.

    b Find the complex number z which satisfies both of these loci.

11 The point P represents the complex number z such that $|z - 1| = 5$

    a Sketch the locus of P and find its cartesian equation.

    b Find the complex numbers represented by the points in this locus which lie on the imaginary axis. Give answers in simplified surd form.

    The point Q represents the complex number z for which

    $\arg(z - 2i) = -\frac{1}{4}\pi$

    c On the same diagram as part **a** sketch the locus of Q and find its cartesian equation.

    d Use algebra to find the complex number which satisfies both loci.

12 P represents the complex number z such that $\left|z-1-i\sqrt{3}\right|=1$

    a Sketch the locus of P.

    b Find the minimum positive value of $\arg z$ for a complex number z in this locus. Give your answer in terms of π.

13 The diagram shows a line l which makes an acute angle of θ radians against the positive x-axis. The equation of l is $y = mx + c$

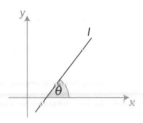

    a Show that $\tan\theta = m$.

    b Hence find the cartesian equation of the locus of points which satisfy $\arg(z-3) = \frac{1}{6}\pi$. Give your answer in the form $\sqrt{a}y + bx + c = 0$ for integers a, b and c to be stated.

14 Point P represents the complex number z where $|z+3+i| = |z-1-i|$

    Point Q represents the complex number z where $|z+5-8i| = \sqrt{5}$

    a Find the cartesian equation of each loci.

    b Use algebra to find the complex numbers z which satisfy both loci.

    c Show that the line AB, where the points A and B represent the complex numbers found in part **b**, is a diameter of the circle defined by the locus of Q.

    d Sketch, on the same diagram, the locus of P and the locus of Q.

15 Sketch, on separate Argand diagrams, these loci.

    a $\arg(3-z) = \frac{1}{4}\pi$ **b** $\arg(2i-z) = \frac{2}{3}\pi$

    c $\arg(1+i-z) = -\frac{1}{6}\pi$ **d** $\arg\left(\frac{1}{z-1-i}\right) = -\frac{1}{2}\pi$

16 Point P represents the complex number z where $|z-2-i| = \sqrt{3}$

    a Sketch the circle C which is the locus of P.

    The point Q represents z where $\arg(z-i) = \theta$, for θ an acute angle. The locus of Q forms a tangent to C.

    b Show that $\theta = \frac{1}{3}\pi$.

    c Find, in the form $a + bi$ for exact $a, b \in \mathbb{R}$, the complex number z which lies in both loci.

17 **a** Shade on the same Argand diagram the region which satisfies both $|z-i| \leqslant 1$ and $\frac{1}{4}\pi \leqslant \arg z \leqslant \frac{1}{3}\pi$

    b Find the exact area of this shaded region.

You can use the geometrical properties of a circle to sketch the locus $\arg\left(\frac{z-a}{z-b}\right) = \theta$ where a and b are complex constants.

In the diagram, points A and B represent fixed complex numbers a and b. Point P represents z, where $\arg\left(\frac{z-a}{z-b}\right) = \theta$ Horizontal lines at A, B and P have been drawn.

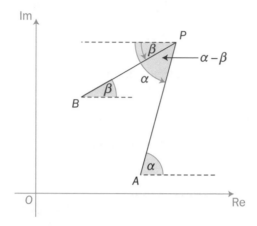

Refer to GCSE: Z-angles.

Vectors $\overrightarrow{AP}$ and $\overrightarrow{BP}$ represent the complex numbers $z - a$ and $z - b$, respectively, where $\alpha = \arg(z - a)$ and $\beta = \arg(z - b)$

Refer to **FP1** Section 2.5.

Using the diagram: angle $APB = \alpha - \beta$

$$= \arg(z - a) - \arg(z - b)$$

$$= \arg\left(\frac{z - a}{z - b}\right)$$

$$= \theta$$

$\arg\left(\frac{z}{w}\right) = \arg z - \arg w$

Hence P moves in such a way that $A\hat{P}B$ is the fixed angle θ. Since angles in the same segment of a circle are equal, the locus of P is an arc of a particular circle, drawn anti-clockwise from A to B.

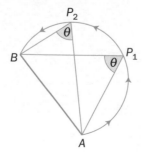

If points A and B represent the complex numbers a and b and P represents the complex number z such that

$$\arg\left(\frac{z-a}{z-b}\right) = \theta, \text{ where } \theta > 0,$$

then the locus of P is an arc of a circle, drawn in an anti-clockwise direction from A to B.

If θ is acute (obtuse) then the locus is the major (minor) arc of the circle.

Points A and B are not included in the locus.

EXAMPLE 1

P represents the complex number z such that
$$\arg\left(\frac{z-1}{z-i}\right) = \frac{4}{5}\pi$$
Sketch the locus of P.

Compare $\arg\left(\frac{z-1}{z-i}\right) = \frac{4}{5}\pi$ with $\arg\left(\frac{z-a}{z-b}\right) = \theta$:

$a = 1 + 0i,\ b = 0 + i,\ \theta = \frac{4}{5}\pi$

Plot the points A and B on the Argand diagram, representing a and b:

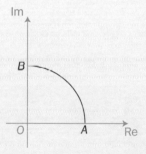

$\theta = \frac{4}{5}\pi$ is obtuse so the locus is the minor arc of a circle.

The locus of P is a minor arc drawn anti-clockwise from A to B.

FP2

You can use an algebraic approach to determine the locus of a point.

EXAMPLE 2

Use algebra to show that the locus of the points which satisfy $|z + 3 + 2i| = 2|z - i|$ is a circle. State the centre and radius of this circle.

Substitute the cartesian form $z = x + iy$ into $|z + 3 + 2i| = 2|z - i|$:

$$|(x + iy) + 3 + 2i| = 2|(x + iy) - i|$$

Collect the real and imaginary parts:

$$|(x + 3) + i(y + 2)| = 2|x + i(y - 1)|$$

so $(x + 3)^2 + (y + 2)^2 = 4(x^2 + (y - 1)^2)$

$|a + ib|^2 = \left(\sqrt{a^2 + b^2}\right)^2 = a^2 + b^2$

Expand each bracket:

$$x^2 + 6x + 9 + y^2 + 4y + 4 = 4(x^2 + y^2 - 2y + 1)$$

Collect like terms:

$$3x^2 - 6x + 3y^2 - 12y - 9 = 0$$

or $\quad x^2 - 2x + y^2 - 4y - 3 = 0$

The coefficients of x^2 and y^2 are the same and there is no term in xy. This is the equation of a circle.

Complete the square in each variable to find the centre and radius of this circle:

$$(x - 1)^2 - 1 + (y - 2)^2 - 4 - 3 = 0$$

so $\quad\quad\quad (x - 1)^2 + (y - 2)^2 = 8$

The locus is a circle, centre $(1, 2)$, radius $\sqrt{8}$.

If P represents the complex number z for which $|z - a| = k|z - b|$, where a and b are fixed complex numbers and $k \in \mathbb{R}$, $k \neq 1$, then the locus of P is a circle.

Exercise 3.6

1 Sketch, on separate Argand diagrams, the locus of points satisfied by

   a $\arg\left(\dfrac{z + 2}{z + 2i}\right) = \dfrac{3}{4}\pi$

   b $\arg\left(\dfrac{z - 3}{z + 3}\right) = \dfrac{1}{3}\pi$

   c $\arg\left(\dfrac{z + 4i}{z}\right) = \dfrac{2}{3}\pi$

   d $\arg\left(\dfrac{z - 2 - i}{z + 4}\right) = \dfrac{1}{2}\pi$

   e $\arg\left(\dfrac{z - 1 + i}{z + 2 - 2i}\right) = \dfrac{1}{5}\pi$

   f $\arg\left(\dfrac{z + 1 + 3i}{z - 4 - i}\right) = 1.6^c$

2 Use algebra to find the cartesian equation of these loci. On separate Argand diagrams, sketch each locus.

   a $|z - 3| = 2|z - 6|$

   b $|z - 1 - 8i| = 3|z - 1|$

   c $|z - 5 + 4i| = 2|z - 2 + i|$

   d $|z - 8 - 6i| = 3|z - 2i|$

   e $|z - 3 + 2i| = \sqrt{2}|z - 1|$

   f $|z + 1 - i| = \sqrt{3}|z - 3 - i|$

3 An arc of a circle C is defined by the locus $\arg\left(\dfrac{z+2}{z-2i}\right)=\dfrac{1}{4}\pi$

   a Show that the point which represents the imaginary number $-2i$ lies in this locus.

   b Sketch the locus of points z which satisfy $\arg\left(\dfrac{z+2}{z-2i}\right)=\dfrac{1}{4}\pi$

   c Write down the complex number represented by the centre of the circle C. Justify your answer.

4 **a** Sketch the locus of points z which satisfy $\arg\left(\dfrac{z-5-2i}{z-1-6i}\right)=\dfrac{1}{2}\pi$

   b Give a complete geometrical description of this locus.

   c Find the greatest value of $|z|$ where z lies in this locus. Give your answer in the form $a+b\sqrt{c}$ for integers a, b and c to be determined.

5 **a** Express $\left|\dfrac{1}{3}z-2+i\right|=\dfrac{\sqrt{5}}{3}|z-2-i|$ in the form $|z-a|=\sqrt{5}|z-b|$ for complex numbers a and b to be stated.

   b Hence, by finding its cartesian equation, sketch the locus of points which satisfy

$$\left|\dfrac{1}{3}z-2+i\right|=\dfrac{\sqrt{5}}{3}|z-2-i|$$

6 The point P represents the complex number z where $|z-1|=2\left|z-\dfrac{1}{4}\right|$

   a Use algebra to show that the locus of P is a circle C centre O and find the radius of this circle.

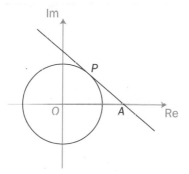

The diagram shows circle C. Point A represents the real number 1. Point P on C is such that the line AP forms a tangent to this circle.

   b Find the exact value of angle PAO.

   c Hence express the locus of this tangent, not including the point A, in the form $\tan(\arg(z-a))=k$ for numbers a and k to be stated.

7 **a** Use an appropriate rule for arguments to show that
$\arg\left(\dfrac{w}{z}\right) = -\arg\left(\dfrac{z}{w}\right)$ where z and w are any non-zero complex numbers.

b Hence sketch, on separate Argand diagrams, the locus of points which satisfy

i $\arg\left(\dfrac{z-3}{z-1}\right) = -\dfrac{2}{3}\pi$ **ii** $\arg\left(\dfrac{z+2-i}{z-3+2i}\right) = -\dfrac{1}{4}\pi$

8 The locus of points P for which $\arg\left(\dfrac{z-7-4i}{z-1-4i}\right) = \dfrac{2}{3}\pi$ is an arc of a circle C.

a Sketch the locus of P on an Argand diagram.

b Explain why the centre of C has real part 4.

c Verify that $4 + \left(4 + \sqrt{3}\right)i$ lies in this locus.

d Hence, or otherwise, find the radius and the centre of C. Give answers in exact form.

9 Use algebra to prove that the locus of points z which satisfy $|z - k^2 c| = k|z - c|$, for $k \neq 1$ and $c = a + bi$ any fixed complex number, is a circle centre O.
Give the radius of this circle in terms of k and $|c|$.

10 On separate Argand diagrams, shade the region defined by these loci.

a $\dfrac{3}{5}\pi \leqslant \arg\left(\dfrac{z-4}{z-1}\right) \leqslant \pi$

b $\dfrac{1}{4}\pi \leqslant \arg\left(\dfrac{z-3+i}{z-3-i}\right) \leqslant \dfrac{1}{2}\pi$

11 The locus of points P such that $\arg\left(\dfrac{z-i}{z+1}\right) = \dfrac{1}{6}\pi$ is an arc of a circle C

a Find

i the complex number λi, where $\lambda > 1$, which lies in this locus,
ii the real number μ, where $\mu < -1$, which lies in this locus.
Give each answer in exact form.

b Sketch the locus of P on an Argand diagram.

c Find, in exact form, the coordinates of the centre of the circle C.

12 The point P represents the complex number z where $\quad |z - 3i| = \sqrt{2}|z - 3|$

a Use algebra to show that the locus of P is a circle.
State the centre and radius of this circle.

b Sketch the locus of P.

c Find, in simplified surd form, the real numbers which lie in this locus.

d i Verify that the complex number $6 + 3i$ lies in the locus of P.
ii Express that part of the locus of P, which consists of the non-real numbers in the first quadrant, in the form

$$\arg\left(\frac{z - a}{z - b}\right) = \theta$$

where $a, b \in \mathbb{R}$ and $0 \leqslant \theta \leqslant \pi$ are in exact form.

13 In the Argand diagram, point P represents the complex number z, where $\quad \arg\left(\frac{z - 3}{z + i}\right) = \frac{1}{2}\pi$

a Verify that the zero complex number lies in the locus of P.

Point Q represents the complex number z where $\quad |z - 3| = |z + i|$

b On a single Argand diagram, sketch the locus of P and the locus of Q.

It is given that $v = \frac{z - 3}{z + i}$, where z is the complex number which belongs to both of these loci.

c State the modulus and argument of v.

d Hence, or otherwise, find z, giving your answer in the form $a + bi$, $a, b \in \mathbb{R}$.

14 The point P in the Argand diagram represents the complex number z where $\quad \arg\left(\frac{z - 1 - i}{z - 2i}\right) = \frac{1}{3}\pi$

a Use a geometrical approach to show that the locus of P does not intersect the imaginary axis.

Use a fact true of all triangles.

b Sketch the locus of P. Mark on your sketch the positions of points A and B which represent the complex numbers $1 + i$ and $2i$ respectively.

c Show that triangle OAB is right-angled.

d Find, in exact form, the modulus and argument of the complex number z in this locus such that $AP = BP$.

You may use the result $(1 + \sqrt{3})^2 = 4 + 2\sqrt{3}$

You can map the complex number $z = x + iy$ to its image $w = u + iv$ by using a transformation.

With this notation, z lies in the z-plane, and its image, w, belongs to the w-plane.

EXAMPLE 1

Under the transformation $w = z^2$

a find the image of $z = 1 + i$

b show that the half-line $\arg z = \frac{1}{4}\pi$ in the z-plane is transformed into a half-line in the w-plane.

a $z = 1 + i$ so $w = z^2 = (1 + i)^2$
$$= 2i$$

So $z = 1 + i$ is mapped to the complex number $w = 2i$

$(1 + i)^2 = 1 + 2i + i^2 = 2i$

b $\arg w = \arg(z^2)$
$$= 2\arg z$$
$$= 2 \times \frac{1}{4}\pi, \text{ or } \frac{1}{2}\pi$$

| Remember that arguments behave like logarithms.

Hence the locus in the w-plane is the half-line $\arg w = \frac{1}{2}\pi$, or the positive section of the imaginary axis (shown in bold in the diagram).

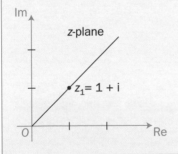

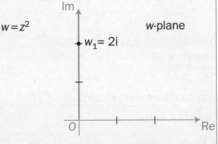

You can use algebra to describe the transformation of a locus.

Use this approach if the equation of the locus in the w-plane produced by the transformation is given in the question.

EXAMPLE 2

The transformation T from the z-plane, where $z = x + iy$, to the w-plane, where $w = u + iv$, is given by $w = \dfrac{z - 2i}{z}$, $z \neq 0$.

Show that T maps the points on the line with equation $y = -x$ in the z-plane, other than $(0,0)$, to points on a straight line in the w-plane with equation $u + v = 1$

Apply the locus equation in the z-plane:

For $z = x + iy$ in the given locus, $y = -x$ so $z = x - ix$

Substitute $z = x - ix$ into $w = \dfrac{z - 2i}{z}$:

$$w = \frac{(x - ix) - 2i}{x - ix}$$

Collect real and imaginary parts:

$$= \frac{x - i(x + 2)}{x(1 - i)}$$

Multiply the numerator and denominator by $(1 - i)^* = 1 + i$:

$$= \frac{(x - i(x + 2))(1 + i)}{2x}$$

Expand the brackets and simplify:

$$= \frac{2x + 2 - 2i}{2x}$$

$(x - i(x + 2))(1 + i)$
$= x + xi - i(x + 2) + x + 2$
$= 2x + 2 - 2i$

Simplify each part:

$$= \frac{x + 1}{x} - \frac{i}{x}$$

Since $w = u + iv$, $\quad u = \dfrac{x + 1}{x}$ and $v = -\dfrac{1}{x}$

Equating real and imaginary parts.

Consider the sum $u + v$:

$$u + v = \frac{x + 1}{x} - \frac{1}{x} = \frac{x + 1 - 1}{x} = \frac{x}{x}$$
$$= 1$$

Let the question guide you.
Work with the sum $u + v$

Hence T maps points on $y = -x$ in the z-plane to points on the line with equation $u + v = 1$ in the w-plane.

You can use modulus properties to determine the transformation of a locus.

Use this approach if the equation of the locus in the w-plane produced by the transformation is not given in the question.

EXAMPLE 3

Show that the transformation given by $w = \dfrac{z}{4 - 2z}$, $z \neq 2$, maps the line $|z| = |z - 4|$ in the z-plane to a line l in the w-plane. Sketch the line l on an Argand diagram.

Rearrange $w = \dfrac{z}{4 - 2z}$ to make z the subject:

$$w = \frac{z}{4 - 2z} \quad \text{so} \quad w(4 - 2z) = z$$

The aim is to replace z in the given locus with an expression involving w.

Multiply out the bracket and collect terms in z:

$$4w = z(2w + 1)$$

Hence $z = \dfrac{4w}{2w + 1}$ where $w \neq -\dfrac{1}{2}$

$w \neq -\dfrac{1}{2}$ avoids division by zero.

Replace z with $\dfrac{4w}{2w + 1}$ in the given locus:

$$|z| = |z - 4| \quad \text{so} \quad \left|\frac{4w}{2w + 1}\right| = \left|\frac{4w}{2w + 1} - 4\right|$$

$$= \left|\frac{4w - 4(2w + 1)}{2w + 1}\right|$$

$$= \left|\frac{-4w - 4}{2w + 1}\right|$$

Rewrite each side of this locus using the rule $\left|\dfrac{a}{b}\right| = \dfrac{|a|}{|b|}$ for $a, b \in \mathbb{R}$:

$$\frac{|4w|}{|2w + 1|} = \frac{|-4w - 4|}{|2w + 1|}$$

Multiply both sides by $|2w + 1|$: $\quad |4w| = |-4w - 4|$

Rewrite each side of this locus, making the coefficient of w equal to 1:

$|4w| = 4|w|$ and $|-4w - 4| = |-4(w + 1)| = 4|w + 1|$

so $\quad 4|w| = 4|w + 1|$

i.e. $\quad |w| = |w + 1|$

Hence l is given by the locus $|w| = |w + 1|$, which is the perpendicular bisector of the line joining $A(0, 0)$ and $B(-1, 0)$.

The locus is $|w| = |w - (-1)|$

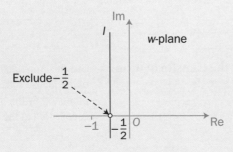

Exclude $-\dfrac{1}{2} + 0\text{i}$ since $w \neq -\dfrac{1}{2}$

The locus consists of all complex numbers of the form $w = -\dfrac{1}{2} + v\text{i}$ provided $v \neq 0$.

Exercise 3.7

1 By expressing z in terms of w, or otherwise, find the locus of points in the w-plane when each of these transformations is applied to the circle $|z - i| = 1$
Sketch, on separate diagrams, each of these loci.

   a $w = z + 2$ b $w = 3 - z$

   c $w = \dfrac{z}{2i}$ d $w = iz$

   e $w = (1 + i)z$ f $w = \dfrac{1}{z}, z \neq 0$

2 For the transformation T given by $w = 2z^3$,

   a find the locus of points in the w-plane when T is applied to these loci:

   i $|z| = 2$ ii $\arg(z) = \dfrac{1}{6}\pi$

   A circle in the z-plane has centre O and radius r. Its image in the w-plane under T is a circle with area $\dfrac{1}{16}\pi$ square units.

   b Find the value of r.

3 By expressing z in terms of w, or otherwise, find the locus of points in the w-plane when each of these transformations is applied to the half-line $\arg(z + 2) = \dfrac{1}{4}\pi$. Sketch, on separate diagrams, each of these loci.

   a $w = \dfrac{1}{2}z$ b $w = 2z + 1$

   c $w = iz - 1$ d $w = \dfrac{1}{z + 2}, z \neq -2$

4 Sketch the locus of points in the w-plane when each of these transformations is applied to the given locus in the z-plane.

   a $|z| = 1, w = \dfrac{z}{z - 1}, z \neq 1$

   b $|z + 2| = 2, w = \dfrac{z + i}{z}, z \neq 0$

   c $|z - i| = |z - 2|, w = \dfrac{iz}{z - i}, z \neq i$

   d $\arg(z) = \dfrac{3}{4}\pi, w = \dfrac{3z + i}{1 - z}, z \neq 1$

5 The transformation T from the z-plane, where $z = x + iy$, to the w-plane, where $w = u + iv$, is given by

   $$w = \dfrac{z}{z + 1}, \quad z \neq -1$$

   Show that T maps the points on the line with equation $y = x + 1$ in the z-plane, other than $(-1, 0)$, to points on a line in the w-plane with equation $u + v = 1$

6 The transformation T from the z-plane, where $z = x + iy$, to the w-plane, where $w = u + iv$, is given by $w = \frac{z}{z-i}$, $z \neq i$.

Show, using two different methods, that T transforms the line $|z - 5i| = |z + 3i|$ in the z-plane to part of a line in the w-plane.

7 The transformation T is given by $w = \frac{3z}{2z-1}$, $z \neq \frac{1}{2}$.

a Show that T maps the circle $\left|z + \frac{1}{2}\right| = 1$ in the z-plane to the line in the w-plane.

T maps the region $\left|z + \frac{1}{2}\right| \geq 1$ in the z-plane to the region R in the w-plane.

b Shade the region R on an Argand diagram.

8 The point P represents the complex number z, where $|2z + 1 + i| = 2|z + 1 + i|$

a Sketch, on an Argand diagram, the locus of P.

The transformation T from the z-plane to the w-plane is given by $w = \frac{z+1}{iz-1}$, $z \neq -i$

b Show that the locus of P under the transformation T is the circle given by $|w - i| = \sqrt{2}|w - 1|$, giving its centre and radius.

9 On separate Argand diagrams, sketch the locus of points in the w-plane when the transformation $w = \frac{z+2i}{z-2}$, $z \neq 2$, is applied to each of these loci in the z-plane.

a $\arg\left(\frac{1}{2}z\right) = \frac{2}{3}\pi$

b $\arg(z + 2i) = \frac{1}{2}\pi$

c $\arg\left(\frac{2z}{z+2i}\right) = \frac{1}{3}\pi$

10 The transformation T given by $w = \frac{z+i}{z}$, $z \neq 0$, maps the z-plane, where $z = x + iy$, to the w-plane, where $w = u + iv$

A point on the line l with equation $y - x + 1 = 0$ in the z-plane is mapped by T to a point $w = u + iv$ in the w-plane.

a Express u and v in terms of x.

b Hence show that $u = (2x - 1)v$

c Show that the image, under T, of the line l is a circle C with cartesian equation $u^2 - u + v^2 - v = 0$

11 The transformation T given by $w = \dfrac{kz}{i+z}$, $z \neq -i$, where k is a real number, maps the complex number $2 + i$ in the z-plane to its image $\frac{1}{2}(3 - i)$ in the w-plane.

a Show that $k = 2$.

Point P represents the complex number z where $|z| = \sqrt{3}$.
T maps point P to point Q in the w-plane.

b Show that the locus of Q is a circle with cartesian equation given by

$$(u - 3)^2 + v^2 = 3 \quad \text{for } u, v \in \mathbb{R}$$

T maps the point z_0 in the locus of P to the point w_0 in the locus of Q, where the acute angle $\arg w_0$ is as large as possible.

c Find the exact value of $|i + z_0|$

12 The point P represents the complex number z, where
$$|z + i| = \sqrt{2}\left|z + \frac{1}{2}i\right|$$

a Show that the locus of P can be expressed as $|z| = \dfrac{1}{\sqrt{2}}$

The point Q represents the complex number z where $\arg z = \frac{1}{4}\pi$
The complex number a lies in both loci.

b Sketch, on the same Argand diagram, the locus of P and the locus of Q.

c Find the complex number a.

The transformation T from the z-plane to the w-plane is given by
$$w = \frac{z + \sqrt{2}i}{\sqrt{2}iz + 1}, \quad z \neq \frac{i}{\sqrt{2}}$$

d Express as a locus in the w-plane the image of P under the transformation T.
Sketch this locus on an Argand diagram.

e Hence, or otherwise, find the imaginary part of the image of the complex number a under T. Give your answer in exact form.

1 Solve the equation $z^4 = -9i$ giving your answers in the form $re^{i\theta}$ where $r > 0$ and $-\pi < \theta \leqslant \pi$ are exact.

2 $z^5 = 4 - 4i$

a Solve this equation, giving your solutions in the form $re^{i\theta}$, where $r > 0$ and $-\pi < \theta \leqslant \pi$ are exact.

b Display these solutions on a single Argand diagram.

c Deduce that $|p| = 2\sin\frac{1}{5}\pi$, where $p = e^{\frac{3}{4}i\pi} - e^{\frac{7}{20}i\pi}$

3 The equation $z^3 + 27i = 0$ has complex roots α, β and γ.

a Find the roots of this equation, giving your answers in exact cartesian form.

b Show algebraically that $\alpha + \beta + \gamma = 0$ and hence find the complex number $(\alpha + \beta)^3$.

c Verify that $-\frac{1}{3}i\alpha$ is a cube root of unity and hence state the value of

$$1 - \frac{1}{3}i\alpha - \frac{1}{9}\alpha^2$$

4 $z = 2 - 2\sqrt{3}i$. Using de Moivre's theorem, or otherwise,

a find z^9, giving your answer in the form a^9 for a an integer to be determined

b evaluate $z^3 - \dfrac{1}{z^3}$

5 a Given that $z = \cos\theta + i\sin\theta$, use de Moivre's theorem to show that

$$z^n + \frac{1}{z^n} = 2\cos n\theta$$

for n any positive integer.

b Hence express $16\cos^4\theta$ in the form $a\cos 4\theta + b\cos 2\theta + c$ where a, b and c are integers.

c Given that $\cos 4\theta = 8\cos^4\theta$, find the two possible values of $\cos\theta$. Give answers in exact form.

FP2

6 a Use de Moivre's theorem to express $\sin 5\theta$ in the form

$$a\cos^4\theta \sin\theta + b\cos^2\theta \sin^3\theta + c\sin^5\theta$$

for integers a, b and c.

b Hence show that, for $\sin\theta \neq 0$,

$$\frac{\sin 5\theta}{\sin^5\theta} \equiv 5\cot^4\theta - 10\cot^2\theta + 1$$

c Solve the equation $\sin 5\theta = 16\sin^5\theta$, for $-\frac{1}{2}\pi \leqslant \theta \leqslant \frac{1}{2}\pi$.
Give your answers in exact form.

7 a Show that $4\sin^3\theta \equiv 3\sin\theta - \sin 3\theta$

b Given that $4\cos^3\theta = \cos 3\theta + 3\cos\theta$, use suitable compound angle formulae to show that

$$32\sin^3\theta\cos^3\theta \equiv 3\sin 2\theta - \sin 6\theta$$

c Verify that the gradient of the curve with equation
$y = \sin^3\theta\cos^3\theta$ at the point where $\theta = \frac{1}{4}\pi$ is zero and
determine the nature of this stationary point.

8 a Use de Moivre's theorem to prove that

$$\cos 6\theta \equiv \cos^6\theta - 15\cos^4\theta\sin^2\theta + 15\cos^2\theta\sin^4\theta - \sin^6\theta$$

and find a similar expression for $\sin 6\theta$.

b Hence show that $\tan 6\theta \equiv 2\tan\theta\left(\dfrac{(3\tan^2\theta - 1)(\tan^2\theta - 3)}{1 - 15\tan^2\theta + 15\tan^4\theta - \tan^6\theta}\right)$
for $\cos 6\theta \neq 0$.

c Given that $\theta = \arctan(\sqrt{2})$ find the exact value of $\tan 6\theta$.

9 a Prove that $64\cos^6\theta = 2\cos 6\theta + 12\cos 4\theta + 30\cos 2\theta + 20$
and find a similar expression for $64\sin^6\theta$.

b Deduce that $\cos^6\theta + \sin^6\theta \equiv \frac{1}{8}(3\cos 4\theta + 5)$

c Hence, or otherwise, state the minimum value of the function
$$f(\theta) = \cos^6\theta + \sin^6\theta, \quad \theta^c \in \mathbb{R}$$
and find in terms of π the smallest positive value of θ which
gives this minimum.

10 a Find the cartesian equation of each of these loci.

   i $|z - 2 - 2i| = \sqrt{2}$ **ii** $\arg(z - 2) = \frac{3}{4}\pi$

b Use algebra to find the complex number which satisfies both loci.

c On the same Argand diagram sketch these loci and state
the geometrical relationship between them.

11 In the Argand diagram the point P represents the complex number z. It is given that $|z - 1| = |z + 1 - i|$

a Show that the locus of P is a straight line and give the cartesian equation of this line.

b Sketch the locus of P.

c On your diagram, shade the region R given by
$$|z - 1| \leqslant |z + 1 - i|$$

d Find the largest value of r such that the circle given by
$$|z - 2| = r$$
lies completely inside the region R. Give your answer in simplified surd form.

12 The point P represents a complex number z on an Argand diagram, where
$$\arg(z + \sqrt{3} + i) = \frac{1}{3}\pi$$

a Given that $z = \lambda i$, where $\lambda \in \mathbb{R}$, is in the locus of P, show that $\lambda = 2$.

The point Q represents a complex number z on an Argand diagram, where
$$|z - 2i| = 2$$

b On the same Argand diagram, sketch the locus of P and the locus of Q.

c Use algebra to find the complex numbers which lie in both of these loci.

13 The point P represents a complex number z on an Argand diagram, where
$$|z - 2 + 3i| = 2|z - 2|$$

a Show that the locus of P is a circle, giving the coordinates of the centre and the radius of this circle.

b Sketch the locus of P.

z_0 is the complex number in the locus of P with greatest negative argument.

c Find $|z_0|$.

14 In the Argand diagram the point P represents the complex number z, where

$$\arg\left(\frac{z-3}{z+i}\right) = \frac{1}{4}\pi$$

Points A and B represent the numbers -1 and $3i$ respectively.

a   Verify that A and B are in the locus of P.

It is given that the locus of P is a major arc of a circle C.

b   Sketch the locus of P. Label the points A and B on your sketch.

c   Find the centre and radius r of C. Give your answer for r in exact form.

15 The point P represents a complex number z on an Argand diagram, where

$$\arg z = \frac{1}{4}\pi$$

a   Sketch the locus of P.

The transformation T from the z-plane to the w-plane is defined by

$$w = iz^2$$

b   Show that T maps the locus of P in the z-plane to the negative part of the real axis in the w-plane.

16 The point P represents a complex number z on an Argand diagram, where

$$|z - 2| = |z - 2i|$$

a   Sketch the locus of P.

The transformation T from the z-plane to the w-plane is defined by

$$w = \frac{z}{z - 2i}, \quad z \neq 2i$$

b   Show that T maps the locus of P in the z-plane to a circle in the w-plane. Give the cartesian equation of this circle.

17 The point P represents a complex number z on an Argand diagram, where

$$|z + 1| = |z + i|$$

a   Sketch the locus of P.

The transformation T from the z-plane to the w-plane is defined by

$$w = \frac{2(1 - z)}{iz - 1}, \quad z \neq -i$$

b   Show that T maps the locus of P in the z-plane to a circle in the w-plane. Give the cartesian equation of this circle.

c   Shade the region in the w-plane which is the image under T of the region in the z-plane given by $|z + 1| \geq |z + i|$

FP2

FP2

Summary

Refer to

- $e^{i\theta} \equiv \cos\theta + i\sin\theta$, $e^{in\theta} + e^{-in\theta} \equiv 2\cos n\theta$, $e^{in\theta} - e^{-in\theta} \equiv 2i\sin n\theta$ 3.1
- You can write any complex number z in the form $z = re^{i\theta}$
  where $r = |z|$, $\theta = \arg z$ 3.1
- For any non-zero complex numbers z and w,

  $|zw| = |z||w|,$ $\left|\dfrac{z}{w}\right| = \dfrac{|z|}{|w|}$

  $\arg(zw) = \arg z + \arg w,$ $\arg\left(\dfrac{z}{w}\right) = \arg z - \arg w$ 3.1
- The equation $z^n = 1$ where $n \in \mathbb{Z}$ has precisely n roots,

  the nth roots of unity, given by $z = e^{\frac{2\pi ki}{n}}$, $k = 0, 1, 2, \ldots, n-1$.
  The sum of these roots is zero.   3.2
- de Moivre's theorem: for any integer n,
  $[r(\cos\theta + i\sin\theta)]^n \equiv r^n(\cos n\theta + i\sin n\theta)$ 3.3
- You can use de Moivre's theorem to prove trigonometric identities.   3.4
- You can sketch the locus of a complex number.   3.5, 3.6
- A transformation is a mapping between the z-plane to the w-plane.   3.7

Links

Many real life phenomena appear to behave
in an unpredictable way.

Mathematicians use complex numbers and
functions to model such situations. This area
of study is known as Chaos theory.

Chaos theory can be used to explain,
for example, the behaviour of weather
systems or populations.

1 a Sketch, on a single diagram, the graphs with equation $y = |3x - 2|$ and the line with equation $y = 4x - 2$

b Solve the inequality $|3x - 2| > 4x - 2$

2 a By expressing $\dfrac{2}{4r^2 - 1}$ in partial fractions, or otherwise, prove that

$$\sum_{r=1}^{n} \frac{2}{4r^2 - 1} = 1 - \frac{1}{2n + 1}$$

b Hence find the exact value of $\displaystyle\sum_{r=11}^{20} \frac{2}{4r^2 - 1}$
[(c) Edexcel Limited 2005]

3 a Solve the equation $z^3 = 4\sqrt{2}(1 + i)$

giving your answers in the form $r(\cos\theta + i\sin\theta)$ where $r > 0$ and $-\pi < \theta \leqslant \pi$ are exact.

b Represent these answers on a single Argand diagram.

c Show that the triangle whose vertices form the solution of the given equation has perimeter $6\sqrt{3}$.

4 a Sketch, on a single diagram, the graphs with equation $y = |x^2 - 3|$ and the line with equation $y = 4x$

b Hence, or otherwise, find the complete set of values of x for which
$$|x^2 - 3| < 4x$$
Give your answers in simplified surd form.

5 a Express $\dfrac{2}{4r^2 + 8r + 3}$ in partial fractions.

b Hence prove that $\displaystyle\sum_{r=1}^{n} \frac{1}{4r^2 + 8r + 3} = \frac{n}{3(2n + 3)}$

c Evaluate $\displaystyle\sum_{r=13}^{39} \frac{1}{4r^2 + 8r + 3}$, giving your answer as a fraction in its simplest form.

6 a Given that $z = e^{i\theta}$, show that
$$z^n - \frac{1}{z^n} = 2i\sin n\theta$$
where n is a positive integer.

b Show that
$$\sin^5\theta = \frac{1}{16}(\sin 5\theta - 5\sin 3\theta + 10\sin\theta)$$

c Hence solve, in the interval $0 \leqslant \theta < 2\pi$,
$$\sin 5\theta - 5\sin 3\theta + 6\sin\theta = 0$$
[(c) Edexcel Limited 2005]

FP2

7 Find the set of values of x such that $|4x - 1| \geqslant |2x - 3|$

8 The point P represents a complex number z on an Argand diagram, where
$$|z - 2 + i| = \sqrt{2}|z - 3|$$

   a Show that the locus of P is a circle, giving the coordinates of the centre and the radius of this circle.

   b Sketch the locus of P.

   c Express the locus of P in the form
$$|z - a| = r \quad \text{where } a \in \mathbb{R} \text{ and } r \in \mathbb{R}.$$

9 a Express as a simplified single fraction $\dfrac{1}{(r - 1)^2} - \dfrac{1}{r^2}$

   b Hence prove, by the method of differences, that
$$\sum_{r=2}^{n} \frac{2r - 1}{r^2(r - 1)^2} = 1 - \frac{1}{n^2}$$
   [(c) Edexcel Limited 2003]

10 Find the set of values of x for which $\dfrac{x^2}{x - 2} > 2x$ [(c) Edexcel Limited 2006]

11 a Show that $(e^{i\theta} - 1)(e^{i\theta} + 1) \equiv 2ie^{i\theta}\sin\theta$

   Given that $z = \dfrac{1}{e^{i\theta} - 1}$, where $\sin\theta \neq 0$

   b **i** find the real part of z

       ii show that the imaginary part of z is $-\frac{1}{2}\cot\left(\frac{1}{2}\theta\right)$.

12 For all real values of r,
$$(2r + 1)^3 - (2r - 1)^3 = Ar^2 + B$$
where A and B are constants.

   a Find the value of A and the value of B.

   b Hence, or otherwise, prove that $\displaystyle\sum_{r=1}^{n} r^2 = \frac{1}{6}n(n + 1)(2n + 1)$ [(c) Edexcel Limited 2006]

13 The transformation T from the z-plane, where $z = x + iy$, to the w-plane, where $w = u + iv$, is given by
$$w = \frac{z + i}{z}, \quad z \neq 0$$

   a The transformation T maps the points on the line with equation $y = x$ in the z-plane, other than $(0,0)$, to points on a line l in the w-plane. Find a cartesian equation of l.

   b Show that the image, under T, of the line with equation $x + y + 1 = 0$ in the z-plane is a circle C in the w-plane, where C has cartesian equation
$$u^2 + v^2 - u + v = 0$$

   c On the same Argand diagram, sketch l and C.   [(c) Edexcel Limited 2007]

14 Solve the inequality $\dfrac{x}{2x+3} \leqslant \dfrac{6}{1-x}$

15 a Express $\dfrac{4}{4r^2 - 4r - 3}$ in partial fractions.

   b Hence show that $\displaystyle\sum_{r=1}^{n} \dfrac{1}{4r^2 - 4r - 3} = \dfrac{n}{1 - 4n^2}$

   c Find an expression, in terms of n, for $\displaystyle\sum_{r=n+1}^{2n^2} \dfrac{1}{4r^2 - 4r - 3}$
   Give your answer in fully factorised form.

16 a Sketch, on a single diagram, the graphs with equation $y = |ax - 2|$ and the line with equation $y = 2x + a$ where $a > 2$ is a constant.

   b Hence, or otherwise, solve the inequality
   $$|ax - 2| < 2x + a$$
   giving your answer in terms of a.

17 a Prove that $\sin^4 \theta \equiv \frac{1}{8}(\cos 4\theta - 4\cos 2\theta + 3)$

   b Hence find the exact value of $\displaystyle\int_0^{\pi} 8\theta \sin^4 \theta \, \mathrm{d}\theta$

18 a Sketch the graph with equation $y = |2x^2 - 5x - 3|$

   b Solve the inequality
   $$|2x^2 - 5x - 3| < 1 - 3x$$
   Give your answer in exact form.

19 In the Argand diagram, the point P represents the complex number z, where
$$\arg\left(\dfrac{z-2}{z+i}\right) = \dfrac{1}{4}\pi$$
The locus of P is an arc A of a circle C.

   a Verify that this locus passes through the point which represents the complex number $0 + 2i$.

   b Find where this locus crosses the negative real axis.

   c Sketch the locus of P.

   d By using an appropriate circle theorem, or otherwise,

   i find the complex number corresponding to the centre of C

   ii show that the arc length A is three-quarters of the circumference of C.

FP2

73

20 a Simplify $r^2(r+1)^2 - (r-1)^2 r^2$

b Hence, or otherwise, prove that

$$\sum_{r=1}^{n} r^3 = \frac{1}{4}n^2(n+1)^2$$

21 a On a single Argand diagram, shade the region R defined by the inequalities

$$\frac{1}{2} < |z| \leqslant 1$$

The transformation T from the z-plane to the w-plane is given by

$$w = \frac{z - 4i}{2z + i}, \quad z \neq -\frac{1}{2}i$$

b Show that the image, under T, of the locus $|z| = 1$ is a circle in the w-plane. Give the centre and radius of this circle.

c Find the cartesian equation of the image, under T, of the locus $|z| = \frac{1}{2}, \quad z \neq -\frac{1}{2}i$

d On a separate Argand diagram, sketch the region in the w-plane which is the image of region R under T.

22 a Express $\dfrac{1}{(r+2)(r+3)}$ in partial fractions.

b Hence show that $\displaystyle\sum_{r=1}^{n} \frac{3}{(r+2)(r+3)} = \frac{n}{n+3}$

c **i** Find an expression in terms of n for the series

$$\sum_{r=1}^{n} \left(2r + \frac{3}{(r+2)(r+3)} \right)$$

giving your answer in fully factorised form.

ii Find the value of $\displaystyle\sum_{r=6}^{21} \left(2r + \frac{3}{(r+2)(r+3)} \right)$

FP2

4

First-order differential equations

This chapter will show you how to
- solve a differential equation by separating its variables
- use boundary conditions to find a particular solution of a differential equation
- formulate a differential equation to model a real-life situation
- solve a linear first-order differential equation by using an integrating factor
- use a substitution to solve a differential equation.

For background knowledge see Sections 0.3 and 0.6 .

Before you start

You should know how to:

1 Solve a differential equation.

e.g. Solve $\dfrac{dy}{dx} = \dfrac{2x}{3y^2}$

Separate the variables and integrate:

$$\int 3y^2 \, dy = \int 2x \, dx$$

i.e. $y^3 = x^2 + c$

The general solution is $y = \sqrt[3]{x^2 + c}$

2 Find and sketch the graph of a particular solution of a differential equation.

e.g. Find and sketch the particular solution of the differential equation $\dfrac{dy}{dx} = 2x + 1$ for which $y = 0$ when $x = 1$.

$\dfrac{dy}{dx} = 2x + 1$ so $y = \displaystyle\int 2x + 1 \, dx = x^2 + x + c$

i.e. $y = x^2 + x + c$

$y = 0$ when $x = 1$, so $0 = 1^2 + 1 + c$, i.e. $c = -2$

Hence $y = x^2 + x - 2 = (x - 1)(x + 2)$

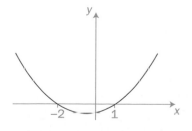

Check in:

See **C4** for revision.

1 Solve these differential equations.

a $\dfrac{dy}{dx} = 2xy^2$

b $\dfrac{dy}{dx} = \dfrac{1}{x\sqrt{y}}$

c $\dfrac{dy}{dx} = \dfrac{x + 1}{2xy}$

2 Find and sketch the graphs of the particular solutions of these differential equations.

a $\dfrac{dy}{d\theta} = y^2 \sin\theta$ over $-\dfrac{1}{2}\pi < \theta < \dfrac{1}{2}\pi$ and for which $y = 1$ when $\theta = 0$.

b $\dfrac{dy}{dx} = \dfrac{x}{y}$ for which $y = 1$ when $x = 1$.

c $\dfrac{dy}{dx} = -\dfrac{y}{x}$ for which $y = 2$ when $x = 1$.

A first-order equation is one which involves a function and its first derivative.

You can solve a **first-order differential equation** by separating its variables and integrating each side.

See **C4** for revision.

The **general solution** of a first-order equation has one arbitrary constant.

EXAMPLE 1

For the differential equation $x\dfrac{dy}{dx} = 3(y - 1)$

a find its general solution, giving your answer in the form $y = f(x)$

b sketch two possible curves with equation $y = f(x)$

---

a Separate the variables ready for integration:

$$x\frac{dy}{dx} = 3(y - 1) \quad \text{so} \quad \int \frac{1}{y-1}\, dy = \int \frac{3}{x}\, dx$$

Leave the constant 3 on the RHS.

Integrate each side: $\quad \ln|y - 1| = 3\ln|x| + c$

Include a constant of integration, c, on one side only.

Replace c with $\ln A$ so that the terms can be combined:

$$\ln|y - 1| = 3\ln|x| + \ln A$$
$$\ln|y - 1| = \ln A|x|^3$$
$$\text{so} \quad |y - 1| = A|x|^3 \quad \text{where } A = e^c > 0$$

Remove the modulus signs by allowing A to take negative values:

$$y - 1 = Ax^3$$

Hence the general solution is $y = Ax^3 + 1$

b The diagram shows the curves with equations $y = 2x^3 + 1$ and $y = -0.1x^3 + 1$

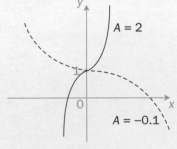

A curve with an equation which satisfies a differential equation is a member of the **family of solution curves** for that differential equation.

You can find the value of c if you are given sufficient information about the equation. Finding c gives a **particular solution** of the differential equation.

EXAMPLE 2

a Find the particular solution of the differential equation
$$x\frac{dy}{dx} = y(1-x)$$
given that $y = \ln 2$ when $x = \ln 2$.

b Sketch the graph of this particular solution for $x \geqslant 0$. You may assume that the curve has exactly one stationary point.

> $x = \ln 2, y = \ln 2$ are **boundary conditions** of the differential equation.

a Separate the variables ready for integration:
$$x\frac{dy}{dx} = y(1-x) \quad \text{so} \quad \int \frac{1}{y}\,dy = \int \frac{1-x}{x}\,dx$$

> You may assume that $x \neq 0$ and $y \neq 0$.

Integrate each side:
$$\ln|y| = \ln|x| - x + c$$

> $$\frac{1-x}{x} = \frac{1}{x} - 1$$

Apply the exponential function to both sides:
$$e^{\ln|y|} = e^{\ln|x|-x+c} = e^{\ln|x|}\,e^{-x}e^{c}$$
$$\text{so} \quad y = Axe^{-x}$$

> Omit the modulus signs by allowing A to be arbitrary.

Substitute the boundary conditions $x = \ln 2, y = \ln 2$ into the general solution to find A:
$$y = Axe^{-x}$$
$$\text{so} \quad \ln 2 = A(\ln 2)(e^{-\ln 2}) = A(\ln 2)\left(\frac{1}{2}\right)$$
$$\text{i.e.} \quad A = 2$$

> $$e^{-\ln 2} = e^{\ln\frac{1}{2}} = \frac{1}{2}$$

Hence the particular solution to the differential equation is $y = 2xe^{-x}$

b The axis-crossing points help determine the general shape of the sketch.

Solve $y = 0$ to find the x-axis crossing points:
$$2xe^{-x} = 0 \text{ has solution } x = 0$$

> $e^{-x} > 0$ for all values of x.

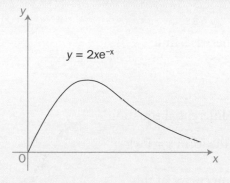

> For large positive values of x the curve $y = 2xe^{-x}$ behaves like the curve $y = e^{-x}$ and so approaches 0 as x increases. See **C3** for revision.

> Unless asked for specific details, you only need to show the basic features of the graph.

If $\lambda > 0$ and n is any positive integer, then for large positive values of x, you can approximate the term $\pm x^{n}e^{-\lambda x}$ by $\pm e^{-\lambda x}$.

> $\approx$ means 'is approximately equal to'.

e.g. For large $x > 0$, $x^{2}e^{-x} \approx e^{-x}$ and $-x^{3}e^{-2x} \approx -e^{-2x}$

FP2

77

Exercise 4.1

1 Find the general solution of these differential equations.
  In each case make y the subject.

  a $x^2 \dfrac{dy}{dx} = \dfrac{2x^2 - 1}{3y^2}$

  b $(2t - 1)\dfrac{dy}{dt} = y(4t^2 - 1)$

  c $\sec\theta \dfrac{dy}{d\theta} = 4\sqrt{y}$

2 Find the particular solution of these differential equations.
  You do not need to make y the subject of each solution.

  a $(2x^2 + 1)\dfrac{dy}{dx} = \dfrac{2x}{y}$ given that $y = 2$ when $x = 0$.

  b $t^2 \dfrac{dy}{dt} = \dfrac{1 - y^3}{y^2}$ given that at $t = 1$, $y = 0$.

  c $(\sqrt{x} + 1)\dfrac{dy}{dx} = \dfrac{y + 1}{\sqrt{x}}$ given that $y = 7$ when $x = 1$.

3 The equation of a curve, defined for $x > 0$, satisfies the
  differential equation

  $$\dfrac{1}{x}\dfrac{dy}{dx} = 4e^{-\frac{1}{2}y}$$

  a Show that the general solution of this equation is given by

  $$y = 2\ln(x^2 + c)$$

  b Given that the curve passes through the point $(1,0)$, find its
    equation and sketch its graph.

4 a Using the substitution $u = x^2 + 1$, or otherwise, find

  $$\int \dfrac{x}{x^2 + 1}\,dx$$

  b Hence find the general solution of the differential equation

  $(x^2 + 1)\dfrac{dy}{dx} = yx$, giving your answer in the form $y = f(x)$

  A curve C is a member of the family of solution curves of this
  differential equation. The point $(0,1)$ lies on C.

  c i Show that the equation of C is given by $y = \sqrt{x^2 + 1}$
    ii Describe the behaviour of C for large positive values of x.

5 a Use a suitable trigonometric identity to show that

$$\int \tan^2 \theta \, d\theta = \tan \theta - \theta + c$$

b Hence find the general solution of the differential equation

$$\cot \theta \frac{dy}{d\theta} = y \tan \theta, \text{ giving } y \text{ in terms of } \theta.$$

6 a Use integration by parts to show that

$$\int 4x \, e^{2x} \, dx = 2x \, e^{2x} - e^{2x} + c$$

b Hence find the particular solution of the differential equation

$$e^{-2x} \frac{dy}{dx} = \frac{4x+1}{y} \quad \text{for which } y = 1 \text{ when } x = 0.$$

Give your answer in the form $y^2 = f(x)$

7 a Express $\dfrac{2}{x^2 - 1}$ in partial fractions.

A member C of the family of solution curves of the differential equation

$$(x - 1)\frac{dy}{dx} = \frac{2y}{x+1} \quad \text{where } x \neq -1$$

passes through the point $P(3, 1)$.

b Find the gradient of C at point P.

c Show that the equation of C may be written as

$$y - 2 = \frac{4}{x+1} \quad \text{and sketch its graph.}$$

8 Obtain the particular solution of the differential equation

$$\sqrt{t^2 - 1} \frac{dy}{dt} = t \cos^2 y \quad \text{where } t > 1$$

and for which $y = \frac{1}{3}\pi$ when $t = 2$.

Give your answer in the form $y = f(t)$

9 a Use integration by parts to show that

$$\int \frac{1}{x^2} \ln x \, dx = -\frac{1}{x}(1 + \ln x) + c$$

A member C of the family of solution curves of the differential equation

$$\frac{dy}{dx} = \frac{y^2 \ln x}{x^2} \quad \text{where } x, y > 0 \text{ passes through the point } P(1, 1).$$

b Show that P is a minimum point on the curve C.

c Show that the equation of C is $y = \dfrac{x}{1 + \ln x}$

d Find, in exact form, the equation of the tangent to C at the point where $x = e$.

e State the value of x for which the curve C is undefined.

FP2

Forming differential equations

EXAMPLE 1

You can formulate a differential equation which models real-life situations.

The population P (in millions) of a particular country t years after 2002 is increasing at a rate which is proportional to the population itself. In 2002 the population was 23.1 million. By 2007, the population had grown to 24.3 million.

a Formulate a differential equation for P.

b Solve your equation from **a**. Give constants to 3 s.f.

c Predict, to the nearest 0.1 million, the population of this country in 2020.

a Use the information provided to formulate the model:

$$\frac{\mathrm{d}P}{\mathrm{d}t} \propto P \quad \text{so} \quad \frac{\mathrm{d}P}{\mathrm{d}t} = kP \quad \text{where } k > 0 \text{ is a constant}$$

$\propto$ means 'is directly proportional to'.
$k > 0$ since the population is increasing.

b Separate the variables and integrate:

$$\int \frac{1}{P}\, \mathrm{d}P = \int k\, \mathrm{d}t \quad \text{so} \quad \ln P = kt + c$$

The modulus symbol is not needed on the LHS, since in this context $P > 0$.

Replace c with $\ln A$ and make P the subject:

$$\ln P = kt + \ln A \Rightarrow P = Ae^{kt}$$

Substitute the boundary conditions, $t = 0$, $P = 23.1$, into this equation to find A:

$$P = Ae^{kt} \quad \text{so} \quad 23.1 = A \times e^{k \times 0}$$
$$\text{i.e.} \quad A = 23.1$$

The year 2002 corresponds to $t = 0$.
P is measured in millions so $P = 23.1$.

Hence $P = 23.1e^{kt}$

$e^0 = 1$

Substitute the boundary conditions, $t = 5$, $P = 24.3$, into this equation to find k:

$$24.3 = 23.1e^{k \times 5} \quad \text{so} \quad 5k = \ln\left(\frac{24.3}{23.1}\right)$$

The year 2007 corresponds to $t = 5$.

Hence $k = \frac{1}{5}\ln\left(\frac{24.3}{23.1}\right) = 0.010\,1287\ldots = 0.0101$ (3 s.f.)

The population can be modelled by the equation $P = 23.1e^{0.0101t}$

c Substitute $t = 18$ into the equation $P = 23.1e^{0.0101t}$:

$$P = 23.1e^{0.0101 \times 18} = 27.705\ldots$$

An estimate for the population in 2020 is 27.7 million, to the nearest 0.1 million.

2020 is 18 years from 2002 so $t = 18$.

It is assumed that the relationship $\frac{\mathrm{d}P}{\mathrm{d}t} \propto P$ continues to hold up to the year 2020.

Exercise 4.2

1 The mass m (in grams) of the radioactive isotope Phosphorus-32 remaining t hours from the start of an experiment decreases at a rate which is directly proportional to the mass present at any time.

   a Express this information as a differential equation.

   It is given that the mass decreases from 20.5 g to 20 g during the first 12 hours.

   b Find m in terms of t. Where appropriate, round constants to 3 significant figures.

   c Solve an appropriate equation to show that it will take approximately 14 days from the start of the experiment for the mass of the substance to halve.

2 A cylindrical barrel (which is initially empty) is being filled with water. After t seconds, the depth of the water in the barrel is h cm. At any instant, the rate at which the depth h is increasing is directly proportional to h. After 20 seconds, the water has a depth of 6.8 cm and is increasing at a rate of $0.17\,\text{cm s}^{-1}$.

   a Formulate a differential equation for h, stating the value of the constant of proportionality.

   b Solve this differential equation to show that h can be modelled approximately by the equation $h = 4.1e^{0.025t}$

   c Given that the container has height 80 cm, use this approximate model to estimate the time taken for the barrel to become full of water.

3 During an experiment, liquid filters into a cylindrical beaker. The beaker, initially empty, has height 20 cm and base radius 5 cm. After t minutes, $V\,\text{cm}^3$ of liquid has filtered into the beaker. At any time the rate at which V is increasing is directly proportional to the volume of space remaining in the beaker at that time.

   a Express this information as a differential equation.

   b Given that after 1 minute the beaker is half-full, show that
   $V = 500\pi(1 - 0.5^t)$

   c **i** Find the time taken for the beaker to be 90% full of liquid. Give your answer in minutes and seconds, correct to the nearest second.

   ii Find the rate at which the volume is increasing at this time. Give your answer to the nearest whole number.

FP2

4   A forest, which initially occupied an area of $25\,\text{km}^2$, is being
destroyed by a fire. The region destroyed by the fire maintains
the shape of a square with side length $L\,\text{km}$.

t hours from the fire beginning, the area A of land (in km^2)
not destroyed satisfies the differential equation $\dfrac{\text{d}A}{\text{d}t} \propto \dfrac{t^3}{L^2}$

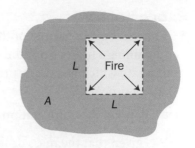

a   Use this information to formulate a differential equation
in terms of A and t only.

After the first 2 hours, $24\,\text{km}^2$ of land remained untouched
by the fire.

b   Show that $A = 25 - \dfrac{t^2}{4}$

After 6 hours, a flash flood completely extinguishes the fire.

c   Calculate the total area of land that was destroyed by the fire
before it was extinguished.

5   In an experiment, a particular chemical is added to a container
of water. The chemical disperses in the shape of a circle. t seconds
from the start of the experiment, the circle has radius $r\,\text{cm}$ and
area $A\,\text{cm}^2$.

a   Assuming that the rate at which A is increasing is inversely
proportional to the radius r at that time, where initially $A = 0$,

　　i   formulate a differential equation in terms of A and hence show
　　　　that
$$A = \sqrt[3]{\lambda t^2} \quad \text{where } \lambda \text{ is a positive constant}$$

　　ii   comment on the validity in the long term of this
　　　　modelling assumption.

b   In a similar experiment, a different chemical also produces a
circular dispersion when added to another container of water.
The variables t, r and A are defined as in the start of the question.
For this chemical, it is assumed that A is increasing at a rate
which is directly proportional to r^A.

　　i   Given that, after 5 seconds, $r = 2\,\text{cm}$ and A is increasing at a
　　　　rate of $16\pi\,\text{cm}^2\,\text{s}^{-1}$, formulate and solve a differential
　　　　equation to show that A can be modelled approximately
　　　　by the equation $A = \dfrac{4\pi}{21 - 4t}$

　　ii   Explain why this model is not appropriate after 6 seconds.

6 The number of sheep, N, on a farm that are infected with an airborne virus t days after its outbreak is increasing in such a way that, at any given time, the rate of infection is directly proportional to the product of the number of infected sheep and non-infected sheep. Before the outbreak, the farm had 150 sheep.

Treat N as a continuous variable.

a Express this information as a differential equation.

At the start of the outbreak, five sheep were infected.

b Show that $\ln\left(\dfrac{N}{150 - N}\right) = \lambda t - \ln 29$

for λ a constant.

After the first day, eight sheep were infected.

c Find the value of λ, giving your answer to 1 decimal place.

d Using this rounded value of λ, show that N can be modelled approximately by the equation

$$N = \frac{150}{1 + 29e^{-0.5t}}$$

e Using this model, estimate
   i the number of sheep which had been infected one week after the outbreak
   ii the rate at which the virus is spreading at this time.

f State the long-term value of N predicted by the model. Comment on whether the answer makes sense in relation to the population of sheep.

FP2

A general **first-order linear differential equation** has the form

$$\frac{dy}{dx} + P(x)y = Q(x) \quad \text{where } P(x) \text{ and } Q(x) \text{ are functions of } x.$$

You can solve a general first-order linear differential equation by multiplying all its terms by an **integrating factor** $I = e^{\int P(x)\,dx}$

The equation is linear since y and $\frac{dy}{dx}$ each occur to the power of 1.

EXAMPLE 1

Find y in terms of x given that $\dfrac{dy}{dx} + \dfrac{y}{x} = 3x$ for $x > 0$.

Compare $\dfrac{dy}{dx} + \dfrac{y}{x} = 3x$ with the general form $\dfrac{dy}{dx} + P(x)y = Q$:

$$P(x) = \frac{1}{x} \quad \text{and} \quad Q(x) = 3x$$

$\dfrac{y}{x} = \dfrac{1}{x}y \quad$ so $\quad P(x) = \dfrac{1}{x}$

Find the integrating factor:

$$I = e^{\int P(x)\,dx} = e^{\int \frac{1}{x}\,dx}$$
$$= e^{\ln x}$$
$$= x$$

You do not need to include the modulus sign or a '+ c' term here.

Multiply all terms in the given differential equation by I:

$$\frac{dy}{dx} + \frac{1}{x}y = 3x \quad \text{so} \quad x\frac{dy}{dx} + y = 3x^2$$

Rewrite the new equation in the form $\dfrac{d(Iy)}{dx} = IQ(x)$:

$$x\frac{dy}{dx} + y = 3x^2 \quad \text{so} \quad \frac{d(xy)}{dx} = 3x^2$$

The integrating factor is such that you can always do this.

$$\frac{d(Iy)}{dx} = \frac{d(xy)}{dx}$$
$$= x\frac{d(y)}{dx} + y\frac{d(x)}{dx}$$
$$= x\frac{dy}{dx} + y$$

Integrate each side with respect to x:

$$xy = x^3 + c$$

Hence the general solution is $y = \dfrac{x^3 + c}{x}$

The general first-order linear differential equation

$\dfrac{dy}{dx} + P(x)y = Q(x)$, where $P(x)$ and $Q(x)$ are functions of x,

can be rewritten as $\quad \dfrac{d(Iy)}{dx} = IQ(x)$

where $I = e^{\int P(x)\,dx}$ is the integrating factor.

The **general solution** y satisfies the equation $Iy = \displaystyle\int IQ(x)\,dx$

You can find the general solution y provided you can integrate the functions $P(x)$ and $IQ(x)$.

You can use integration techniques from C4 to help you solve more complicated differential equations.

EXAMPLE 2

a Use integration by parts to show that

$$\int xe^x \, dx = xe^x - e^x + c$$

b Hence find the particular solution of the differential equation $x\dfrac{dy}{dx} + 2y = e^x$

for $x \neq 0$ and for which $y = 2$ when $x = 1$.

---

a Use the integration by parts formula:

$$\int u\frac{dv}{dx}\,dx = uv - \int v\frac{du}{dx}\,dx$$

so $\displaystyle\int xe^x\,dx = xe^x - \int e^x\,dx$

$$= xe^x - e^x + c \quad \text{as required.}$$

$u = x \Rightarrow \dfrac{du}{dx} = 1$

$\dfrac{dv}{dx} = e^x \Rightarrow v = \displaystyle\int e^x\,dx = e^x$

b Express the given equation in the form $\dfrac{dy}{dx} + P(x)y = Q(x)$:

$$x\frac{dy}{dx} + 2y = e^x$$

Divide through by x: $\quad \dfrac{dy}{dx} + \dfrac{2}{x}y = \dfrac{e^x}{x}$

Make sure you divide both sides by x.

Find the integrating factor:

$$I = e^{\int P(x)\,dx} = e^{\int \frac{2}{x}\,dx}$$

$$= e^{2\ln x}$$

$$= e^{\ln x^2}$$

$$= x^2$$

$P(x) = \dfrac{2}{x}, \; Q(x) = \dfrac{e^x}{x}$

Use the result $Iy = \displaystyle\int IQ(x)\,dx$ to find y:

$$x^2 y = \int xe^x\,dx$$

$IQ(x) = x^2 \times \dfrac{e^x}{x} = xe^x$

Use the result from part **a**: $\quad x^2 y = xe^x - e^x + c$

Hence the general solution is given by $y = \dfrac{e^x(x-1) + c}{x^2}$

Use the boundary conditions $x = 1$, $y = 2$ to find c:

$$y = \frac{e^x(x-1) + c}{x^2}$$

$$2 = \frac{e^1(1-1) + c}{1^2} \quad \text{so } c = 2$$

Hence the particular solution is $y = \dfrac{e^x(x-1) + 2}{x^2}$

Exercise 4.3

1 Use an integrating factor to find the general solution of these differential equations.

a $\dfrac{dy}{dx} + \dfrac{y}{x+1} = 3x - 3$

b $\dfrac{dy}{dx} + \dfrac{2y}{x} = \dfrac{1}{x^2}$

c $\dfrac{dy}{dx} + \dfrac{y}{x-2} = \dfrac{1}{x}$

d $(x+2)\dfrac{dy}{dx} + 2y = 3x$

e $x^2\dfrac{dy}{dx} - xy = x^3 - 1$

f $\theta\dfrac{dy}{d\theta} + 2y = \cos(\theta^2)$

g $\dfrac{2}{t}\dfrac{dx}{dt} + \dfrac{x}{t^2} = 2\sqrt{t}$

h $\dfrac{1}{(x+1)}\dfrac{dy}{dx} + \dfrac{y}{x} = 6e^{-x}$

i $x\dfrac{dy}{dx} + (x^2 - 1)y = x^3$

2 Given that $x\dfrac{dy}{dx} + (1+x)y = e^{-x}$

a show that xe^x is an integrating factor for this differential equation

b find the particular solution of this differential equation for which

$$y = \tfrac{1}{e} \text{ when } x = 1$$

c sketch the graph of this solution.

3 a Solve the differential equation $\dfrac{dy}{dx} + 4xy = 2x$

i by using an integrating factor
ii by separating the variables.

b Account for any apparent difference in the two expressions obtained.

4 a Express $\dfrac{1}{x(x+1)}$ in partial fractions.

b Hence find the general solution to the differential equation

$$(x+1)\dfrac{dy}{dx} + \dfrac{y}{x} = \dfrac{x+1}{x}$$

5 $\dfrac{dy}{dx} - y\tan x = x$

a Find $\displaystyle\int x\cos x\,dx$

b Show that $\cos x$ is an integrating factor of this differential equation.

c Find the particular solution of this differential equation given that $y = 0$ when $x = \pi$.

6 **a** Use a suitable trigonometric identity to show that

$$\int \sin\theta \sin 2\theta \, d\theta = \frac{2}{3}\sin^3\theta + c$$

b Hence find the general solution to the differential equation

$$\frac{dy}{d\theta} + y\cot\theta = 3\sin 2\theta$$

c Show that the particular solution of the given equation which satisfies the condition $y = 1$ at $\theta = \frac{1}{4}\pi$ is given by $y = 2\sin^2\theta$

d Sketch the graph of this solution.

7 **a** Using a suitable substitution, or otherwise, find

$$\int \frac{1}{(x+1)^2} \, dx$$

b Hence find the general solution to the differential equation

$$(x+1)\frac{dy}{dx} + xy = e^{-x}$$

8 Find the general solution of the differential equation

$$\sin x \frac{dy}{dx} + (\cos x - \sin x)y = e^{2x}$$

9 **a** Prove that $2\operatorname{cosec} 2x \equiv \cot x + \tan x$

b Hence find the general solution of the differential equation

$$\sin 2x \frac{dy}{dx} + 2y = Q(x)$$

in each of the following cases.
i $Q(x) = 2\cos^2 x$
ii $Q(x) = 2\cos 2x$

You can simplify a differential equation which cannot be solved easily by using a **change of variable** (or **substitution**).

EXAMPLE 1

Given that y satisfies the differential equation

$$x^2\frac{dy}{dx} - xy = y^2 \text{ for } x > 0 \qquad (1)$$

a show that the substitution $y = ux$, where u is a function of x, transforms equation (1) into

$$x\frac{du}{dx} = u^2 \qquad (2)$$

b find the general solution of equation (2)

c hence write down the general solution of equation (1).

> In the exam, the substitution will be given in the question.

---

a $y = ux$ so $\dfrac{dy}{dx} = \dfrac{d(ux)}{dx} = u + x\dfrac{du}{dx}$

> Use the product rule. You can assume that u is a function of x only.

Substitute for y and $\dfrac{dy}{dx}$ in equation (1):

$$x^2\frac{dy}{dx} - xy = y^2 \qquad (1)$$

> Do not assume that u is a constant.

> $y^2 = (ux)^2 = u^2x^2$

so $x^2\left(u + x\dfrac{du}{dx}\right) - x^2u = u^2x^2$

> LHS $= x^2u + x^3\dfrac{du}{dx} - x^2u$
> $\qquad = x^3\dfrac{du}{dx}$

Cancel the x^2u terms:

$$x^3\frac{du}{dx} = u^2x^2$$

i.e. $\qquad x\dfrac{du}{dx} = u^2$

> You can divide through by x^2, as $x \neq 0$.

giving equation (2), as required.

b Separate the variables of equation (2) and integrate:

$x\dfrac{du}{dx} = u^2$ so $\displaystyle\int\frac{1}{u^2}\,du = \int\frac{1}{x}dx$

$$\text{i.e. } -\frac{1}{u} = \ln x + c$$

Hence the general solution to equation (2) is $u = \dfrac{-1}{(\ln x + c)}$

> The modulus sign on x is not needed, since $x > 0$.

c Since $y = ux$, the general solution of equation (1) is given by

$$y = \frac{-x}{(\ln x + c)}$$

> Use the solution for u in **b** to find the solution for y.

Exercise 4.4

1 In each question

   i use the substitution $y = ux$ to find and solve a differential equation for u

   ii find y in terms of x.

   a $x\dfrac{dy}{dx} - y = 2xy^2$

   b $xy\dfrac{dy}{dx} - y^2 = 3x^4$

   c $xy^2\dfrac{dy}{dx} - y^3 = x^2$

   d $x^2\dfrac{dy}{dx} - y^2 = xy$

   e $x\dfrac{dy}{dx} - y = x\sqrt{y}$

2 $x\dfrac{dy}{dx} = y(1 - ye^x)$ for $x > 0$ (1)

   a Show that the substitution $y = ux$ transforms the given differential equation into the equation

   $$\dfrac{du}{dx} = -u^2 e^x \qquad (2)$$

   b Find the general solution of equation (2).

   c Hence find the particular solution of equation (1) for which $y = \dfrac{1}{e}$ when $x = 1$.

   Sketch the graph of this solution for $x > 0$.

3 For the differential equation $\dfrac{dy}{dx} - \dfrac{3y}{2x} = \dfrac{x}{2y}$, where $x, y > 0$,

   a show that the substitution $y = ux$ transforms the given differential equation into the equation

   $$\dfrac{du}{dx} = \dfrac{(u^2 + 1)}{2u}x$$

   b solve this differential equation for u and hence find y in terms of x.

4 Use the substitution $y = ux$ to obtain the general solution of each differential equation. Give your answers in the form $y^2 = f(x)$

   a $x\dfrac{dy}{dx} - y = \dfrac{x^3\cos x}{2y}$

   b $x\dfrac{dy}{dx} - y = \dfrac{x^3\cot x}{2y}$

   c $\dfrac{x^2}{y}\dfrac{dy}{dx} = (y\sec x)^2 + x$

5 A curve C is defined over the interval $0 < x \leqslant \pi$.
   The equation of C satisfies the differential equation

$$xy\frac{dy}{dx} - y^2 = x^3 \sin 2x \qquad (1)$$

   a Show that the substitution $y = ux$ transforms equation (1)
      into the equation

$$u\frac{du}{dx} = \sin 2x \qquad (2)$$

   b Find the general solution of equation (2).

   c i   Given that the curve passes through the point $\left(\frac{1}{4}\pi, \frac{1}{4}\pi\right)$

          show that the equation of C is $y = \sqrt{2}x\sin x$

      ii Find the exact coordinates of the other point where C
          intersects the line $y = x$

6 a Given that $v = yx$, where y is a function of x, show that
      $$\frac{dv}{dx} = x\frac{dy}{dx} + y$$

   b Use the substitution $v = yx$ to show that the differential equation

$$x\frac{dy}{dx} + y = 2x^3y^2 \qquad (1)$$

      can be re-expressed as

$$\frac{dv}{dx} = 2v^2x \qquad (2)$$

   c Find the general solution of equation (2) and hence show that
      $y = \dfrac{1}{x(c - x^2)}$ for c an arbitrary constant.

7 a Express $\dfrac{4}{u^2 - 4}$ in partial fractions.
      The function $y(x)$ satisfies the differential equation

$$x\left(\frac{dy}{dx} + 1\right) - y = \frac{y^2 - 3x^2}{x} \qquad (1) \qquad \text{where } x > 0$$

   b Show that the substitution $y = ux$ transforms equation (1)
      into the differential equation

$$\frac{du}{dx} = \frac{u^2 - 4}{x} \qquad (2)$$

   c Show that the general solution of equation (2) is given by
      $u = 2\left(\dfrac{kx + 1}{1 - kx}\right)$ for k an arbitrary constant.

   d Find the particular solution of equation (1) given that $y = -3$ at $x = 3$.

8 a Show that the substitution $y = ux$ transforms the differential equation

$$(y - 2x)\frac{dy}{dx} = 2y - x \qquad (1)$$

into the differential equation

$$\frac{du}{dx} = \frac{u^2 - 4u + 1}{x(2 - u)} \qquad (2)$$

b Find the general solution of equation (2).

c Hence show that the general solution of equation (1) satisfies the equation

$$y = 2x \pm \sqrt{A + 3x^2} \quad \text{for } A \text{ an arbitrary constant.}$$

d A particular solution of equation (1) is such that $y = 4$ when $x = 1$.

i Find the equation of this particular solution in the form $y = f(x)$

ii Deduce that large values of x, $f(x)$ can be approximated by the expression $(2 + \sqrt{3})x$

9 a Find $\displaystyle\int \frac{1}{x \ln x}\,dx$

b Use the substitution $y = ux$ to show that the differential equation

$$x\frac{dy}{dx} = y(\ln y - \ln x + 1) \quad \text{where } x, y \geqslant 0 \qquad (1)$$

can be transformed into the equation

$$\frac{du}{dx} = \frac{u \ln u}{x} \qquad (2)$$

c Find the general solution of equation (2).

d i Find the particular solution of equation (1) for which

$$y = \frac{1}{e^2} \quad \text{when } x = 1$$

ii Sketch the graph of this particular solution for $x \geqslant 0$.
You may assume this curve has exactly one stationary point.

1 The equation of a curve C satisfies the differential equation

$$\frac{dy}{dx} = \frac{2xy}{x^2 + 1}$$

 a Find the general solution of this equation.

 b Sketch, on one diagram, two possible graphs of the curve C.

2 a Use integration by parts to find $\int te^t \, dt$

 b Hence show that the particular solution of the differential equation

$$\frac{dx}{dt} = te^{t-x}, \quad t > 1$$

where $x = 2$ when $t = 2$, can be expressed as $x = t + \ln(t-1)$

3 The behaviour of oil slicks on the sea is being modelled in a large water tank. At the start of one trial, oil is continuously released into the water and begins to spread out in a circle, centred at the point of release. After t hours, the slick has radius r metres and depth h metres, where $h = \frac{1}{100}(1 - 0.1t)$

At all times, the radius is increasing at a rate which is inversely proportional to the volume (in m³) of oil released at that point in time.

 a Express this information as a differential equation for r.

One hour after the start of the experiment, the slick has radius 2 metres.

 b Show that r can be approximately modelled by the equation

$$r^3 = -76 \ln(100h)$$

 c Use this approximation to estimate the total time required for the slick to have a radius of 4 metres.
  Give your answer in hours to 1 decimal place.

 d Explain why the model is not appropriate 10 hours after the start of the trial.

4 The equation of a curve C satisfies the differential equation

$$x\frac{dy}{dx} - y = x^3$$

 a Find the general solution of this equation.

 b Given that C passes through the point $(2,3)$ find the equation of C.
  Sketch its graph, marking all axis crossing points with their coordinates.

5 a Find $\displaystyle\int \frac{x}{x+1}dx$

b Hence show that the particular solution of the differential equation

$$(x+1)\frac{dy}{dx} + xy = \frac{x+1}{xe^x}, \quad x > 0$$

for which $y = 0$ when $x = 1$ is given by $y = \dfrac{(x+1)}{e^x}\ln\!\left(\dfrac{2x}{x+1}\right)$

6 a Use a suitable double-angle formula to show that

$$\int \cos^2\theta \, d\theta = \tfrac{1}{2}\sin 2\theta + \theta + c$$

b For the differential equation $\dfrac{dx}{d\theta} + x\tan\theta = 4\cos^3\theta$

   i show that $\sec\theta$ is an integrating factor for this equation

   ii find the general solution of this equation.

7 a Show that the substitution $y = ux$ transforms the differential equation

$$xy\frac{dy}{dx} = 2y^2 - x^2, \quad x > 0 \qquad (1)$$

into the equation

$$\frac{du}{dx} = \frac{u^2 - 1}{ux} \qquad (2)$$

b Find the general solution of equation (2).

c Hence find the general solution of equation (1).
Give your answer in the form $y^2 = f(x)$

8 The equation of a curve C satisfies the differential equation

$$x\frac{dy}{dx} + 2x^2 y = y(2x + 1), \quad x > 0, \, y > 0$$

The point with coordinates $(2, 6)$ lies on C.

a Use the substitution $y = vx$ to form and solve a differential equation in v.

b Hence show that the equation of C is given by $y = 3xe^{x(2-x)}$

c Find the equation of the tangent to the curve at the point $x = 1$. Give your answer in terms of e.

d Find the exact x-coordinate of the stationary point of C.

FP2

FP2

Summary

Refer to

- You can solve a differential equation $\dfrac{dy}{dx} = f(x)g(y)$,

  where $f(x)$ is a function of x only and $g(y)$ is a function of y only, by separating its variables and integrating each side.

  4.1

- The rate of change of the quantity A is $\dfrac{dA}{dt}$, where the variable t represents time.

  4.2

- The general solution of the first-order linear differential equation $\dfrac{dy}{dx} + P(x)y = Q(x)$, where $P(x)$ and $Q(x)$ are functions of x, satisfies the equation

  $$Iy = \int IQ(x)\,dx, \quad \text{where } I = e^{\int P(x)\,dx} \text{ is an integrating factor.}$$

  4.3

- You can simplify a differential equation by using a change of variable.

  4.4

Links

Differential equations and their solutions are used extensively in physical sciences.

Newton's law of cooling, which describes how a substance cools over time, is a first-order differential equation given by

$$\frac{dT}{dt} = k(T - T_0)$$

where, in appropriate units, T is the temperature of the substance after time t, k is a constant of proportionality and T_0 is the (constant) surrounding temperature.

5

Second-order differential equations

This chapter will show you how to
- solve a general second-order differential equation
- use boundary conditions to find a particular solution of a differential equation
- use a substitution to solve a differential equation.

For background knowledge see
Sections 0.3, 0.5, ◯, and **FP1**.

FP2

Before you start

You should know how to:

1 Solve a quadratic equation.

e.g. Solve $x^2 + 2x + 5 = 0$

Complete the square: $(x + 1)^2 - 1 + 5 = 0$

so $(x + 1)^2 + 4 = 0$

$(x + 1)^2 = -4$

Take square roots: $x + 1 = \pm 2i$

Hence the roots are $x = -1 \pm 2i$

2 Find the second derivative of a function.

e.g. If $y = 2xe^{3x}$ find $\dfrac{d^2y}{dx^2}$

$y = 2xe^{3x}$ so $\dfrac{dy}{dx} = 2x(3e^{3x}) + e^{3x}(2)$

$= 2e^{3x}(3x + 1)$

Hence $\dfrac{d^2y}{dx^2} = 2e^{3x}(3) + (3x + 1)(6e^{3x})$

$= 6e^{3x}(3x + 2)$

3 Approximate a function for large values of x.

e.g. State the approximate value of
$g(x) = x^2e^{-x} + 3$ for large positive values of x.
For large positive values of x, $x^2e^{-x} \approx 0$.
Hence $g(x) \approx 3$ for large positive values of x.

Check in:

1 Solve

a $2x^2 - 2x + 1 = 0$

b $3x^2 + 4x + 1 = 0$

c $2m^2 + 2m + 5 = 0$

d $4m^2 + 12m + 9 = 0$

See **FP1** Chapter 2.

2 Find $\dfrac{d^2y}{dx^2}$ when

a $y = 3e^x + 4e^{2x}$

b $y = x\sin 2x$

c $y = e^{2x}(2x + 1)$

d $y = x(\cos 2x + \sin 2x)$

See **C3** for revision.

3 Find the approximate value of these functions for large positive values of x.

a $g(x) = x(2e^{-x} + 1) - x$

b $h(x) = (xe^{-x} + 2)^2$

The equation $a\dfrac{d^2y}{dx^2} + b\dfrac{dy}{dx} + cy = 0$

You can solve the second-order differential equation

$$a\frac{d^2y}{dx^2} + b\frac{dy}{dx} + cy = 0$$

by first solving the quadratic equation $am^2 + bm + c = 0$

> The equation is second order because the highest derivative in the equation is $\dfrac{d^2y}{dx^2}$.
>
> a, b and c are constants.

$am^2 + bm + c = 0$ is known as the auxiliary equation.

You can determine the general solution of the differential equation $\quad a\dfrac{d^2y}{dx^2} + b\dfrac{dy}{dx} + cy = 0$

from the nature of the roots α and β of the auxiliary equation:

> The general solution is also known as the complementary function.

	Nature of roots	The general solution (A, B arbitrary constants)
Case 1	α, β real, $\alpha \neq \beta$	$y = Ae^{\alpha x} + Be^{\beta x}$
Case 2	α, β real, $\alpha = \beta$	$y = e^{\alpha x}(Ax + B)$
Case 3	α, β complex, $p \pm iq$	$y = e^{px}(A\cos qx + B\sin qx)$

> In each case, you need two arbitrary constants because the equation is second order.
>
> In Case 2, $\alpha = \beta$ means α is a repeated root.
>
> In Case 3, complex roots must be conjugate pairs (see **FP1**).

EXAMPLE 1 (CASE 1)

a Find the general solution of the differential equation

$$\frac{d^2y}{dx^2} - 5\frac{dy}{dx} + 6y = 0$$

b Verify that your answer to **a** is the general solution of the given equation.

---

a Compare $a\dfrac{d^2y}{dx^2} + b\dfrac{dy}{dx} + cy = 0$ with the given equation

$\dfrac{d^2y}{dx^2} - 5\dfrac{dy}{dx} + 6y = 0$:

$a = 1, b = -5, c = 6$

Write down and solve the auxiliary equation to find α and β:

$$m^2 - 5m + 6 = 0$$
$$(m - 3)(m - 2) = 0$$

so $\qquad \alpha = 3, \beta = 2$

> Use α and β rather than the letter m for the roots.

Write down the general solution using the roots $\alpha = 3$, $\beta = 2$:

The general solution is $y = Ae^{3x} + Be^{2x}$

> Case 1 applies since the roots are real and distinct.

EXAMPLE 1 (CONT.)

b Differentiate to find $\frac{dy}{dx}$ and $\frac{d^2y}{dx^2}$:

$$y = Ae^{3x} + Be^{2x} \quad \text{so} \quad \frac{dy}{dx} = 3Ae^{3x} + 2Be^{2x}$$

$$\text{and} \quad \frac{d^2y}{dx^2} = 9Ae^{3x} + 4Be^{2x}$$

$$\frac{d(e^{kx})}{dx} = ke^{kx}$$

Substitute these expressions into the left-hand side of the equation:

$$\frac{d^2y}{dx^2} - 5\frac{dy}{dx} + 6y$$

$$= (9Ae^{3x} + 4Be^{2x}) - 5(3Ae^{3x} + 2Be^{2x}) + 6(Ae^{3x} + Be^{2x})$$

Collect the terms in e^{3x} and e^{2x}.

$$= (9A - 15A + 6A)e^{3x} + (4B - 10B + 6B)e^{2x}$$
$$= 0 \times e^{3x} + 0 \times e^{2x}$$
$$= 0$$

The answer 0 confirms that $y = Ae^{3x} + Be^{2x}$ is the general solution of the differential equation $\frac{d^2y}{dx^2} - 5\frac{dy}{dx} + 6y = 0$

EXAMPLE 2 (CASE 2)

a Find the complementary function of the differential equation $4\frac{d^2y}{dx^2} + 4\frac{dy}{dx} + y = 0$

b Determine the behaviour of this function for large values of x.

---

a Write down and solve the auxiliary equation:
$$4m^2 + 4m + 1 = 0$$

Complete the square: $\quad (2m + 1)^2 = 0$

i.e. $\qquad\qquad\qquad \alpha = -\frac{1}{2}$

The auxiliary equation has a repeated root $\alpha = -\frac{1}{2}$.

Hence, by Case 2, the complementary function is
$$y = e^{-\frac{1}{2}x}(Ax + B)$$

b In the expression $e^{-\frac{1}{2}x}(Ax + B)$, the exponential term dominates.

For large values of x, $e^{-\frac{1}{2}x} \approx 0$

and so $y = e^{-\frac{1}{2}x}(Ax + B) \approx 0$

$4\frac{d^2y}{dx^2} + 4\frac{dy}{dx} + y = 0$
so $a = 4, b = 4, c = 1$

There is a repeated root so Case 2 applies.

For large x, $e^{-\frac{1}{2}x}$ is the dominant term in the solution. Refer to Section 4.1.

FP2

EXAMPLE 3 (CASE 3)

Find y in terms of x given that $\dfrac{d^2y}{dx^2} + 4\dfrac{dy}{dx} + 5y = 0$

Write down and solve the auxiliary equation:
$$m^2 + 4m + 5 = 0$$

Complete the square to find its roots:
$(m+2)^2 - 4 + 5 = 0$ so $(m+2)^2 = -1$
i.e. $m = -2 \pm i$

$-2 \pm i$ is of the form $p \pm iq$, where $p = -2$ and $q = 1$.

Hence the general solution is $y = e^{-2x}(A\cos x + B\sin x)$

$\dfrac{d^2y}{dx^2} + 4\dfrac{dy}{dx} + 5y = 0$

so $a = 1, b = 4, c = 5$

The roots are complex so Case 3 applies.

You would obtain the same solution by letting $q = -1$ due to the properties of $\sin x$ and $\cos x$.

Exercise 5.1

1 Find the general solutions of these differential equations. In each case check by substitution that your answer is correct.

a $\dfrac{d^2y}{dx^2} - 4\dfrac{dy}{dx} - 5y = 0$

b $\dfrac{d^2y}{dx^2} + 5\dfrac{dy}{dx} + 6y = 0$

c $\dfrac{d^2y}{dx^2} - 4\dfrac{dy}{dx} + 4y = 0$

d $2\dfrac{d^2y}{dx^2} - 3\dfrac{dy}{dx} - 2y = 0$

e $\dfrac{d^2y}{dx^2} - 4\dfrac{dy}{dx} + 5y = 0$

f $\dfrac{d^2y}{dx^2} + 8\dfrac{dy}{dx} + 16y = 0$

g $4\dfrac{d^2y}{dx^2} - 4\dfrac{dy}{dx} + y = 0$

h $5\dfrac{d^2y}{dx^2} + 4\dfrac{dy}{dx} + y = 0$

i $5\dfrac{d^2y}{dx^2} + 6\dfrac{dy}{dx} + 5y = 0$

2 Find the complementary function for each differential equation.

a $\dfrac{d^2u}{dx^2} + 7\dfrac{du}{dx} + 12u = 0$ **b** $4\dfrac{d^2x}{dt^2} + 12\dfrac{dx}{dt} + 9x = 0$ **c** $\dfrac{d^2y}{dx^2} + 9y = 0$

3 a Find y in terms of x given that $3\dfrac{d^2y}{dx^2} + 2\dfrac{dy}{dx} = 0$

b Describe the behaviour of y as x increases $(x > 0)$.

4 Find the complementary function of each differential equation.

a $\dfrac{d^2y}{dx^2} + 6\dfrac{dy}{dx} + 9y = 0$ **b** $\dfrac{d^2x}{dt^2} + 6\dfrac{dx}{dt} + 13x = 0$ **c** $4\dfrac{d^2y}{dt^2} - 3\dfrac{dy}{dt} - y = 0$

d $4\dfrac{d^2y}{dx^2} + 9y = 0$ **e** $3\dfrac{d^2u}{dx^2} + 4\dfrac{du}{dx} - 4u = 0$ **f** $9\dfrac{d^2v}{dt^2} - 6\dfrac{dv}{dt} + v = 0$

5 Consider the differential equation $a\dfrac{d^2y}{dx^2} - 2\dfrac{dy}{dx} + 3y = 0$,
where $a \neq 0$ is a constant.

  a Given that $y = e^{3x}$ satisfies this equation, show that $a = \dfrac{1}{3}$.

  b Hence find the general solution of this equation.

6 a Show that the complementary function for the equation

$$\dfrac{d^2y}{dx^2} - 4\dfrac{dy}{dx} + 20y = 0$$

   is given by $y = g(x)$, where $g(x) = e^{2x}(A\cos 4x + B\sin 4x)$ for
   A and B arbitrary constants.

  b For any given pair of values for A and B, where $A \neq 0$, find the

   exact value of $\dfrac{g(\pi)}{g(0)}$.

7 a Find the general solution y of the differential equation

$$\dfrac{d^2y}{dx^2} - 10\dfrac{dy}{dx} + 25y = 0$$

  b Hence find the differential equation of the form $\dfrac{d^2y}{dx^2} + b\dfrac{dy}{dx} + cy = 0$
   of which $y - e^{-2x}$ is a solution.

8 a Show that if $y = e^x\cos x$ then $\dfrac{dy}{dx} - e^x(\cos x - \sin x)$

   and find an expression for $\dfrac{d^2y}{dx^2}$ in terms of x.

   It is given that $y = e^x\cos x$ satisfies the differential equation

$$\dfrac{d^2y}{dx^2} - 2\dfrac{dy}{dx} + ky = 0, \quad \text{where } k \text{ is a constant}$$

  b Show that $k = 2$.

  c Hence find the general solution to this equation.

9 The equation of a curve C satisfies the differential equation

$$\dfrac{d^2y}{dx^2} - 2k\dfrac{dy}{dx} + k^2y = 0, \quad \text{where } k \text{ is a non-zero constant.}$$

  a Show that the general solution of this equation is given by
   $y = e^{kx}(Ax + B)$, for A and B arbitrary constants.

   For a given pair of non-zero values of A and B, the point $P(x_0, y_0)$
   where $y_0 \neq 0$, is a stationary point of the curve C.

  b i Express x_0 in terms of k, A and B.
   ii Show that the nature of this stationary point depends
      only on the sign of y_0 and clearly state this relationship.

FP2

To solve the equation $a\dfrac{d^2y}{dx^2} + b\dfrac{dy}{dx} + cy = f(x)$,

where $f(x)$ is *not* the zero function:

Step 1 Find the complementary function (CF) of

$$a\dfrac{d^2y}{dx^2} + b\dfrac{dy}{dx} + cy = 0$$

Step 2 Find a function y_T of x which satisfies the

equation $a\dfrac{d^2y}{dx^2} + b\dfrac{dy}{dx} + cy = f(x)$

y_T is the particular integral (PI) of the
differential equation.

Step 3 Add the PI to the CF to give the
general solution (GS) $y = \text{CF} + \text{PI}$

y_T is also called a **trial function**.

The particular integral depends on $f(x)$:

	$f(x)$ (lower case constants given)	Particular integral y_T (upper case constants to be found)
Type 1 $f(x)$ is exponential	$f(x) = de^{kx}$	Try $y_T = De^{kx}$
Type 2 $f(x)$ is trigonometric	$f(x) = d\cos qx + e\sin qx$ (including $d = 0$ or $e = 0$)	Try $y_T = D\cos qx + E\sin qx$
Type 3 $f(x)$ is a polynomial, degree $\leqslant 2$	$f(x) = d,\ dx + e$ or $dx^2 + ex + f$	Try $y_T = D,\ Dx + E$ or $Dx^2 + Ex + F$ respectively

EXAMPLE 1

Find a particular integral of the differential equation
$\dfrac{d^2y}{dx^2} - 3\dfrac{dy}{dx} + 2y = f(x)$ in the case when

a $f(x) = x^2 + x - 1$ **b** $f(x) = 2\sin x$

---

a $f(x) = x^2 + x - 1$ is a polynomial (Type 3).

Use the table to write down an appropriate particular integral y_T:
$y_T = Dx^2 + Ex + F$

You want to find the value of
the constants D, E and F.

Find the first and second derivatives of y_T in terms of D, E and F:
If $y = Dx^2 + Ex + F$ then

$$\dfrac{dy}{dx} = 2Dx + E \quad \text{and} \quad \dfrac{d^2y}{dx^2} = 2D$$

Substitute these expressions into the given differential equation:

$$\frac{d^2y}{dx^2} - 3\frac{dy}{dx} + 2y = x^2 + x - 1$$

so $\quad 2D - 3(2Dx + E) + 2(Dx^2 + Ex + F) = x^2 + x - 1$

Simplify the LHS:

$$(2D)x^2 + (2E - 6D)x + (2D - 3E + 2F) = x^2 + x - 1$$

Compare coefficients of like terms:

x^2 terms: $\quad 2D = 1$ so $D = \frac{1}{2}$

x terms: $\quad 2E - 6D = 1$ so $E = 2$

constants: $\quad 2D - 3E + 2F = -1$, so $F = 2$

Hence $D = \frac{1}{2}, E = 2, F = 2$

A particular integral of this differential equation is

$$y_T = \frac{1}{2}x^2 + 2x + 2$$

b $f(x) = 2\sin x$ is a trigonometric function (Type 2).

Use the table to write down an appropriate particular integral y_T:

$$y_T = D\cos x + E\sin x$$

Find the first and second derivatives of y_T in terms of D, E and F:

If $y = D\cos x + E\sin x$ then

$$\frac{dy}{dx} = -D\sin x + E\cos x$$

and $\quad \frac{d^2y}{dx^2} = -D\cos x - E\sin x$

Substitute these expressions into the given differential equation:

$$\frac{d^2y}{dx^2} - 3\frac{dy}{dx} + 2y = 2\sin x \quad \text{so}$$

$$(-D\cos x - E\sin x) - 3(-D\sin x + E\cos x) + 2(D\cos x + E\sin x) = 2\sin x$$

Simplify the LHS by collecting terms in $\cos x$ and in $\sin x$:

$$(D - 3E)\cos x + (3D + E)\sin x = 2\sin x$$

Hence $\quad D - 3E = 0 \quad$ (1)

and $\quad 3D + E = 2 \quad$ (2)

Solve equations (1) and (2) simultaneously:

Substituting $D = 3E$ from (1) into (2) gives $10E = 2$

so $\quad D = \frac{3}{5}, E = \frac{1}{5}$

Hence a particular integral is $y_T = \frac{3}{5}\cos x + \frac{1}{5}\sin x$

By definition, y_T satisfies the given differential equation.

The x^2 coefficient on the LHS is $2D$ and on the RHS is 1.

Substituting $D = \frac{1}{2}$

Substituting $D = \frac{1}{2}, E = 2$

$$y_T = Dx^2 + Ex + F$$

You must include $\cos x$ in y_T, even though $f(x)$ depends only on $\sin x$.

This relationship must hold for *all* values of x.

The RHS does not involve $\cos x$ so $D - 3E = 0$

FP2

You can decide if a given function is a particular integral of a differential equation by substituting its derivatives into the equation.

EXAMPLE 2

Given that $\dfrac{d^2y}{dx^2} + 3\dfrac{dy}{dx} + 2y = 2e^{-x}$

a verify that $2xe^{-x}$ is a particular integral of this differential equation

b find the general solution of this differential equation.

a Find the first and second derivatives of $y_T = 2xe^{-x}$:
$$y = 2xe^{-x}$$

Use the product rule:
$$\frac{dy}{dx} = 2(1-x)e^{-x} \quad \text{and} \quad \frac{d^2y}{dx^2} = -2(2-x)e^{-x}$$

Substitute these expressions into the LHS of the differential equation:
$$\frac{d^2y}{dx^2} + 3\frac{dy}{dx} + 2y = -2(2-x)e^{-x} + 3\big(2(1-x)e^{-x}\big) + 2(2xe^{-x})$$
$$= (-2(2-x) + 6(1-x) + 4x)e^{-x}$$
$$= (-4 + 2x + 6 - 6x + 4x)e^{-x}$$
$$= 2e^{-x}$$

Hence $2xe^{-x}$ is a particular integral of $\dfrac{d^2y}{dx^2} + 3\dfrac{dy}{dx} + 2y = 2e^{-x}$

b Find the complementary function:
$$\frac{d^2y}{dx^2} + 3\frac{dy}{dx} + 2y = 0 \text{ has CF } y = Ae^{-x} + Be^{-2x}$$

Use GS = CF + PI:

$y_T = 2xe^{-x}$ is a particular integral if it satisfies the given differential equation.

This is the RHS of the differential equation.

The auxiliary equation is $m^2 + 3m + 2 = 0$ so $(m+1)(m+2) = 0$ i.e. the roots are $\alpha = -1$, $\beta = -2$.

In Example 2, for the equation $\dfrac{d^2y}{dx^2} + 3\dfrac{dy}{dx} + 2y = 2e^{-x}$, a particular integral was of the form Dxe^{-x} rather than just De^{-x}. This is because both the complementary function and f(x) involve a common term (e^{-x}).

CF $= Ae^{-x} + Be^{-2x}$, f(x) $= 2e^{-x}$

If, in a differential equation, the complementary function and f(x) share a common term then you can find an appropriate particular integral by multiplying y_T by x (or, if this does not work, by x^2) where y_T is the usual trial function.

Exam questions usually give the appropriate PI in this case.

You can check if y_T is correct by using the method shown in part **a** of Example 2.

Exercise 5.2

1 Find a particular integral of each of these differential equations.

a $\dfrac{d^2y}{dx^2} + 3\dfrac{dy}{dx} + 2y = 2e^x$

b $2\dfrac{d^2y}{dx^2} - \dfrac{dy}{dx} - y = 5e^{-3x}$

c $\dfrac{d^2y}{dx^2} - \dfrac{dy}{dx} - 2y = 20\sin 2x$

d $2\dfrac{d^2y}{dx^2} + 6\dfrac{dy}{dx} - y = 5\cos 2x - 10\sin 2x$

e $\dfrac{d^2y}{dx^2} + \dfrac{dy}{dx} - 2y = 3x - 2$

f $\dfrac{d^2y}{dx^2} + 2\dfrac{dy}{dx} - 8y = 4x^2 + 2$

2 Find the general solution of these differential equations.

a $\dfrac{d^3y}{dx^2} - 4\dfrac{dy}{dx} + 4y = 3e^{-x}$

b $\dfrac{d^2y}{dt^2} - \dfrac{dy}{dt} - 2y = 9t$

c $4\dfrac{d^2y}{dt^2} + 4\dfrac{dy}{dt} + y - \cos t - 7\sin t$

d $\dfrac{d^2y}{dx^2} - 4\dfrac{dy}{dx} + 5y = 6e^{3x}$

e $\dfrac{d^2x}{dt^2} - 2\dfrac{dx}{dt} + 10x = 5t^2 + 8t - 21$

f $\dfrac{d^2y}{d\theta^2} + 2\dfrac{dy}{d\theta} + 5y - \cos 4\theta - 6\sin 4\theta$

3 Given that y satisfies the differential equation

$$\dfrac{d^2y}{dx^2} + 8\dfrac{dy}{dx} + 16y = 7e^{-3x}$$

a find y in terms of x

b determine the behaviour of y as x increases, for $x > 0$.

4 The equation of a curve C satisfies the differential equation

$$\dfrac{d^2y}{dx^2} + 4\dfrac{dy}{dx} + 4y = 8\cos 2x$$

a Find the general solution of this differential equation.

b Sketch, for large positive values of x, a curve which approximates the graph of C.

5 $\dfrac{d^2y}{dx^2} + \dfrac{dy}{dx} - 6y = 5e^{2x}$

a Verify that $2xe^{2x}$ is a particular integral of this differential equation.

b Hence find the general solution of this differential equation.

6 $\dfrac{d^2y}{dx^2} - 3\dfrac{dy}{dx} = 3x - 1$

 a Given that λx^2, where λ is a constant, is a particular integral
   of this differential equation, show that $\lambda = -\dfrac{1}{2}$.

 b Hence find the general solution of this differential equation.

7 $\dfrac{d^2y}{dx^2} + 4y = 4\sin 2x$

 a Find the value of the constant k such that $kx\cos 2x$ is a
   particular integral of this differential equation.

 b Hence find the general solution of this differential equation.

8 The equation $y(x)$ of a curve C satisfies the differential equation

   $2\dfrac{d^2y}{dx^2} + 5\dfrac{dy}{dx} + 2y = 6x + 3$

 a Find the general solution of this differential equation.

 b Hence show that for large positive values of x, the graph of C may be
   approximated by a straight line. State the equation of this line.

9 The equation $y(x)$ of a curve C satisfies the differential equation

   $\dfrac{d^2y}{d\theta^2} + 5\dfrac{dy}{d\theta} + 6y = 4\sin\theta$

 a Find the general solution of this differential equation.

 b Hence show that, for large positive values of θ, the equation
   of C may be approximated by a sine function and write
   down this function.
   Show further that, for large positive values of θ, $\left(\dfrac{dy}{d\theta}\right)^2 + y^2 \approx \dfrac{8}{25}$

10 By finding a suitable particular integral, find the general solution
   y of the differential equation $\dfrac{d^2y}{dx^2} + 2\dfrac{dy}{dx} = f(x)$ when

 a $f(x) = 3e^{-2x}$

 b $f(x) = 1 - x^2$

   In each case, describe the behaviour of y for large positive values of x.

11 a Show that the general solution of the differential equation

$$9\frac{d^2y}{dx^2} + 6\frac{dy}{dx} + y = 2 - 12x - x^2$$

can be expressed as $y = (A + Bx)e^{-\frac{1}{3}x} - x^2 + 20$
for arbitrary constants A and B.

A particular member C of the family of solution curves of this equation is such that $A = -5$.

b i Find the y-intercept of C.
ii Deduce that C crosses the positive real axis.
You should not attempt to find where this happens.

12 a Show that the general solution of the differential equation

$$\frac{d^2y}{dx^2} + 6\frac{dy}{dx} + 10y = 15\cos x + 3\sin x$$

is given by $y = g(x)$, where
$$g(x) = e^{-3x}(A\cos x + B\sin x) + \cos x + \sin x$$

for A and B arbitrary constants.

b Sketch the general shape of the graph with equation

$$y = (g(x))^2 \quad \text{for } 10\pi \leqslant x \leqslant 12\pi$$

You may assume that the exponential term is approximately zero over this interval.

c Find a cosine function which approximates $g(x)$ for large positive values of x.

FP2

With sufficient information, you can find the value of the two arbitrary constants in the general solution of a second-order differential equation.

Compare with the first-order case in Section 4.1.

EXAMPLE 1

Find the particular solution of the differential equation

$$\frac{d^2y}{dx^2} - 2\frac{dy}{dx} - 3y = 0$$

for which $y = 1$ and $\frac{dy}{dx} = -5$ when $x = 0$.

Solve the auxiliary equation:

$m^2 - 2m - 3 = 0$ has roots $\alpha = -1$, $\beta = 3$

The complementary function is $y = Ae^{-x} + Be^{3x}$

Substitute the boundary conditions $x = 0$, $y = 1$ into the CF:

$$y = Ae^{-x} + Be^{3x}$$
$$1 = Ae^{-0} + Be^{3\times0}$$

so $1 = A + B$ \hfill (1)

Differentiate the CF to find a second equation in A and B:

$$y = Ae^{-x} + Be^{3x}$$

so $\dfrac{dy}{dx} = -Ae^{-x} + 3Be^{3x}$

Substitute the boundary conditions $x = 0$, $\dfrac{dy}{dx} = -5$ into this equation:

$$\frac{dy}{dx} = -Ae^{-x} + 3Be^{3x}$$

so $-5 = -A + 3B$ \hfill (2)

Solve the simultaneous equations:

$$A + B = 1 \qquad (1)$$
$$-A + 3B = -5 \qquad (2)$$

(1) + (2) gives $A = 2$ and $B = -1$

Hence the required particular solution is $y = 2e^{-x} - e^{3x}$

EXAMPLE 2

The equation of a curve C satisfies the differential equation

$$\frac{d^2y}{dx^2} + 2\frac{dy}{dx} + y = 2x$$

The curve passes through the origin and the point $P(-4, -12)$.

a Find the equation of C.

b Hence show that for large positive values of x, C may be approximated by a straight line. State the equation of this line.

---

a The equation of C is the particular solution of the given differential equation satisfying the boundary conditions $x = 0, y = 0$ and $x = -4, y = -12$.

> Point $P(-4, -12)$ is on C so when $x = -4$, $y = -12$

Solve the auxiliary equation:

$m^2 + 2m + 1 = 0$ has a repeated root $\alpha = -1$.

Hence the complementary function is $y = e^{-x}(Ax + B)$

Find a particular integral:

$f(x) = 2x$ is a linear function so a PI is of the form $y_1 = Dx + E$

> Refer to Section 5.2.

Find the first and second derivatives of the PI to find D and E:

If $y = Dx + E$ then $\frac{dy}{dx} = D$ and $\frac{d^2y}{dx^2} = 0$

Substitute these expressions into the given differential equation:

$$\frac{d^2y}{dx^2} + 2\frac{dy}{dx} + y = 2x \quad \text{so} \quad 0 + 2D + (Dx + E) = 2x$$

$$Dx + (2D + E) = 2x$$

Compare x-coefficients: $D = 2$

Compare constant terms: $2D + E = 0$ so $E = -4$

Hence a particular integral is $y_T = 2x - 4$

So the general solution of the differential equation is

$$y = e^{-x}(Ax + B) + 2x - 4$$

> $y_T = Dx + E = 2x - 4$

Substitute $x = 0, y = 0$ into the general solution to find an equation in A and B:

$$0 = e^{-0}(A \times 0 + B) + 0 - 4 \quad \text{so} \quad B = 4$$

Substitute $x = -4$ and $y = -12$ into the general solution to find a second equation in A and B:

$$-12 = e^4(-4A + 4) - 12$$

$$e^4(-4A + 4) = 0$$

So $A = 1$

Hence the equation of curve C is $y = e^{-x}(x + 4) + 2x - 4$

> When finding A, replace B with 4 in the general solution.
>
> $e^4 \neq 0$ so $-4A + 4 = 0$

b For large positive values of x, $e^{-x}(x + 4) \approx 0$

Hence the straight line with equation $g(x) = 2x - 4$ provides a close approximation to the curve C.

Exercise 5.3

1 Use the given boundary conditions to find the particular solution of these differential equations.

a $\dfrac{d^2y}{dx^2} - 7\dfrac{dy}{dx} + 10y = 0$, given that $y = 4$ and $\dfrac{dy}{dx} = 11$ when $x = 0$.

b $4\dfrac{d^2y}{dx^2} - 12\dfrac{dy}{dx} + 9y = 0$, given that $y = 4$ and $\dfrac{dy}{dx} = 3$ when $x = 0$.

c $9\dfrac{d^2y}{dx^2} + 12\dfrac{dy}{dx} + 4y = 0$, given that at $x = 0$, $y = 5$ and $\dfrac{dy}{dx} = -\dfrac{4}{3}$.

d $\dfrac{d^2y}{dt^2} + 4\dfrac{dy}{dt} + 13y = 0$, given that $y = 2$ and $\dfrac{dy}{dt} = -7$ when $t = 0$.

e $\dfrac{d^2x}{d\theta^2} - 4\dfrac{dx}{d\theta} + 8x = 0$, given that $x = 2$ when $\theta = 0$, and

$x = e^{\frac{1}{2}\pi}$ when $\theta = \dfrac{1}{4}\pi$.

2 Find the general solution of these differential equations.

a $2\dfrac{d^2y}{dx^2} - \dfrac{dy}{dx} - y = x^2 + 1$, given that $y = -8$ and $\dfrac{dy}{dx} = -5$ when $x = 0$.

b $4\dfrac{d^2y}{dx^2} + y = 10e^x$, given that when $x = 0$, $y = 5$ and when $x = \pi$, $y = 2e^\pi$.

c $\dfrac{d^2y}{d\theta^2} + 6\dfrac{dy}{d\theta} + 9y = -7\cos\theta - \sin\theta$, given that $y = \dfrac{3}{2}$ and $\dfrac{dy}{d\theta} = -\dfrac{7}{2}$ when $\theta = 0$.

3 The equation $y(x)$ of a curve C satisfies the differential equation

$$\dfrac{d^2y}{dx^2} - y = 3e^{2x}$$

The curve passes through the origin and the point $P(\ln 2, 5)$.

a Show that the equation of C is given by $y = e^x - 2e^{-x} + e^{2x}$

b Hence find the value of y when $x = \ln\left(\dfrac{1}{2}\right)$

4 **a** Find the particular solution of the differential equation

$$\dfrac{d^2y}{dx^2} - 2\dfrac{dy}{dx} + 5y = \cos x + 3\sin x$$

for which $y = \dfrac{3}{2}$ when $x = 0$, and $y = e^{\frac{3\pi}{4}}$ when $x = \dfrac{3}{4}\pi$.

b Find the exact value of y when $x = \pi$.

5 A function $y(x)$ satisfies the differential equation

$$4\frac{d^2y}{dx^2} - 4\frac{dy}{dx} + y = -\frac{1}{4}x^2 + 3x - 6$$

a Find y in terms of x given that $y = 1$ and $\frac{dy}{dx} = 1.25$ when $x = 0$.

b Hence show that the equation $y = 0$ has exactly one positive root. State the value of this root.

6 The equation of a curve C satisfies the differential equation

$$\frac{d^2y}{dx^2} - 6\frac{dy}{dx} + 10y = 3\cos 2x + 6\sin 2x$$

The equation of the tangent to this curve at its y-intercept is $y = x + 1$

a Show that the equation of C is given by $y = \frac{1}{2}(\cos 2x + e^{3x}(\cos x - \sin x))$

b Verify that the graph of C crosses the x-axis at $x = \frac{1}{4}\pi$ and find the exact gradient of C at this point.

7 It is given that $kt\sin 4t$, where k is a constant, is a particular integral of the differential equation $\frac{d^2y}{dt^2} + 16y = 4\cos 4t$

a Show that $k = \frac{1}{2}$.

b Hence find the particular solution of this differential equation for which $y = 2$ when $t = 0$, and $y = 1$ when $t = \frac{1}{8}\pi$.

8 The equation of a curve C satisfies the differential equation

$$\frac{d^2y}{dx^2} - 2\frac{dy}{dx} = 4e^x$$

The curve passes through the point with coordinates $(0, 1)$.
The gradient of the curve at this point is 4.

a Show that the equation of C can be expressed as $y = (2e^x - 1)^2$

b Find the exact coordinates of the point P where this curve crosses the x-axis.

c i Show that point P is a minimum point.
   ii Sketch the graph of C, showing clearly the behaviour of the curve for large positive and large negative values of x.

You can simplify a second-order differential equation which is difficult to solve by using a **change of variable** (or **substitution**).

Compare with Section 4.4.

EXAMPLE 1

For the differential equation

$$x^2 \frac{d^2y}{dx^2} - 2x\frac{dy}{dx} + 2(1 + 2x^2)y = 0, \quad \text{where } x \neq 0 \qquad (1)$$

a show that the substitution $y = ux$ transforms equation (1) into

$$\frac{d^2u}{dx^2} + 4u = 0 \qquad (2)$$

b by solving equation (2), find the general solution of equation (1).

In the exam you will be given the required substitution.

---

a Find $\frac{dy}{dx}$ in terms of u and x: $y = ux$ (*)

Apply the product rule to ux: $\dfrac{dy}{dx} = x\dfrac{du}{dx} + u$ (**)

Find an expression for $\frac{d^2y}{dx^2}$:

$$\frac{d^2y}{dx^2} = \frac{d}{dx}\left(\frac{dy}{dx}\right) = \frac{d}{dx}\left(x\frac{du}{dx} + u\right)$$

Apply the product rule to $x\dfrac{du}{dx}$:

$$= x\frac{d^2u}{dx^2} + \frac{du}{dx} + \frac{du}{dx}$$

$$= x\frac{d^2u}{dx^2} + 2\frac{du}{dx} \qquad (***)$$

So $y = ux$, $y' = xu' + u$ and $y'' = xu'' + 2u'$

Replace y, y' and y'' in (1) with these expressions:

$$x^2 y'' - 2xy' + 2(1 + 2x^2)y = 0 \qquad (1)$$

so $x^2(xu'' + 2u') - 2x(xu' + u) + 2(1 + 2x^2)xu = 0$

i.e. $x^3 u'' + 4x^3 u = 0$

Divide both sides by x^3:

$$u'' + 4u = 0$$

i.e. $\dfrac{d^2u}{dx^2} + 4u = 0,$ which is equation (2), as required.

b The differential equation $\dfrac{d^2u}{dx^2} + 4u = 0$ has general

solution $u = A\cos 2x + B\sin 2x$

Since $y = ux$, the general solution of equation (1) is

$y = x(A\cos 2x + B\sin 2x)$

You can assume that u is a function of x only. Do not assume that u is constant.

Using (**)

Using dashed notation to simplify the working (where

$u' = \dfrac{du}{dx}, y' = \dfrac{dy}{dx}, y'' = \dfrac{d^2y}{dx^2}$, etc):

$y' = xu' + u$ (**)
$y'' = (xu'' + u') + u'$
   $= xu'' + 2u'$ (***)

The terms in $x^2 u'$ cancel as do the terms in xu.

$x^3 \neq 0$ since $x \neq 0$

The auxiliary equation $m^2 + 4 = 0$ has roots $\pm 2i$.

Exercise 5.4

1 Use the substitution $y = ux$, where u is a function of x, to solve these differential equations, giving y in terms of x.

a $\quad x^2 \dfrac{d^2y}{dx^2} - 2x\dfrac{dy}{dx} + 2(1 - 2x^2)y = 0$

b $\quad x^2 \dfrac{d^2y}{dx^2} + x(x - 2)\dfrac{dy}{dx} + (2 - x)y = 0$

c $\quad x^2 \dfrac{d^2y}{dx^2} - 2x(2x + 1)\dfrac{dy}{dx} + 2(1 + 2x + 4x^2)y = 0$

d $\quad 4x^2\dfrac{d^2y}{dx^2} + 4x(3x - 2)\dfrac{dy}{dx} + (8 - 12x + 9x^2)y = 0$

2 a Show that the substitution $y = ux$ transforms the differential equation

$$x^2\dfrac{d^2y}{dx^2} + 2x(x - 1)\dfrac{dy}{dx} + (x^2 - 2x + 2)y = 0 \qquad (1)$$

into the equation

$$\dfrac{d^2u}{dx^2} + 2\dfrac{du}{dx} + u = 0 \qquad (2)$$

b Solve equation (2), giving u in terms of x and hence write down the general solution of equation (1).

3 a Show that the substitution $y = ux$ transforms the differential equation

$$4x^2\dfrac{d^2y}{dx^2} - 8x\dfrac{dy}{dx} + (8 + x^2)y = x^5 + 8x^3 \qquad (1)$$

into the equation

$$4\dfrac{d^2u}{dx^2} + u = x^2 + 8 \qquad (2)$$

b Find the general solution of equation (2).

c Hence show that the general solution of equation (1) is given by

$$y = x\left(A\cos\tfrac{1}{2}x + B\sin\tfrac{1}{2}x + x^2\right)$$

for arbitrary constants A and B.

FP2

4 The equation of a curve C satisfies the differential equation

$$x^2\frac{d^2y}{dx^2} + x(x-2)\frac{dy}{dx} + \left(\frac{1}{4}x^2 - x + 2\right)y = 0 \qquad (1)$$

a Show that the substitution $y = vx$ transforms the equation (1) into the differential equation

$$4\frac{d^2v}{dx^2} + 4\frac{dv}{dx} + v = 0 \qquad (2)$$

b Solve equation (2), giving v in terms of x.

c Given that the curve passes through the point $(1,0)$ and that $\frac{dy}{dx} = 1$ when $x = 0$

   i   find the equation of C
   ii sketch the graph of C for $x \geqslant 0$.

5 A curve C is a member of the family of solution curves of the differential equation

$$x^2\frac{d^2y}{dx^2} - 2x(x+1)\frac{dy}{dx} + 2(x+1)y = 6x^3e^x \qquad (1)$$

a Show that the substitution $y = xv$ transforms equation (1) into the differential equation

$$\frac{d^2v}{dx^2} - 2\frac{dv}{dx} = 6e^x \qquad (2)$$

b Solve equation (2).

c Given that C crosses the x-axis when $x = \ln 2$ and passes through the point $(\ln 3, -\ln 3)$ show that the equation of C is given by
$$y = x(e^{2x} - 6e^x + 8)$$

d Find the exact values of x where the curve crosses the positive x-axis again.

6 a Given that $y = v - x$, where v is a function of x, find expressions for $\frac{dy}{dx}$ and $\frac{d^2y}{dx^2}$ in terms of v.

b Find the general solution of the differential equation

$$4\frac{d^2y}{dx^2} - 4\frac{dy}{dx} + y = 4 - x$$

   i   directly
   ii by using the substitution $y = v - x$

7 The equation of a curve C satisfies the differential equation

$$x^2\frac{d^2y}{dx^2} - 2x\frac{dy}{dx} + 2(1 + 2x^2)y + 4x^3 = 0 \qquad (1)$$

a Show that the substitution $y = ux$ transforms equation (1) into the differential equation

$$\frac{d^2u}{dx^2} + 4u = -4 \qquad (2)$$

b Solve equation (2) and hence write down the general solution of equation (1).

The curve passes through the point (π, π) and has gradient -3 when $x = \frac{1}{2}\pi$.

c Show that the equation of C is given by $y = x(2\cos 2x - 1)$ and find the exact value of y when $x = \frac{1}{2}\pi$.

d Sketch the graph of C over the interval $0 \leqslant x \leqslant \pi$.

8 a If $v = xy$, where y is a function of x, show that $\dfrac{dv}{dx} = x\dfrac{dy}{dx} + y$ and find an expression for $\dfrac{d^2v}{dx^2}$.

b Show that the substitution $v = xy$ transforms the differential equation

$$x\frac{d^2y}{dx^2} + (2 + 3x)\frac{dy}{dx} + (3 + 2x)y = 0 \qquad (1)$$

into the equation

$$\frac{d^2v}{dx^2} + 3\frac{dv}{dx} + 2v = 0 \qquad (2)$$

c Solve equation (2) and hence write down the general solution of equation (1).

9 a If $y = \dfrac{1}{x}$, where x is a function of t, show that $\dfrac{dy}{dt} = -\dfrac{1}{x^2}\dfrac{dx}{dt}$ and find an expression for $\dfrac{d^2y}{dt^2}$.

b Show that the substitution $y = \dfrac{1}{x}$ transforms the differential equation

$$x\frac{d^2x}{dt^2} - 2\frac{dx}{dt}\left(x + \frac{dx}{dt}\right) = 0 \qquad (1)$$

into the equation

$$\frac{d^2y}{dt^2} - 2\frac{dy}{dt} = 0 \qquad (2)$$

c Solve equation (2), giving y in terms of t.

d i Find the particular solution $x(t)$ of equation (1) for which $x = 1$ when $t = 0$ and $x = \frac{4}{7}$ when $t = \ln 2$.

ii Deduce that $x(t) < \frac{4}{3}$ for all values of t.

FP2

1 Find the general solution of each differential equation.

a $\dfrac{d^2y}{dx^2} + \dfrac{dy}{dx} - 20y = 0$

b $\dfrac{d^2y}{dx^2} + \dfrac{2}{3}\dfrac{dy}{dx} + \dfrac{1}{9}y = 0$

c $\dfrac{d^2x}{dt^2} - 8\dfrac{dx}{dt} + 17x = 0$

2 a Find the general solution of the differential equation

$$3\dfrac{d^2y}{dx^2} + 7\dfrac{dy}{dx} - 6y = 2e^{-2x}$$

b i Hence show that the particular solution of this differential equation

for which $y = 0$ and $\dfrac{dy}{dx} = -3$ when $x = 0$ is given by

$$y = -\dfrac{3}{4}e^{\frac{2}{3}x} + e^{-3x} - \dfrac{1}{4}e^{-2x}$$

ii Describe the behaviour of this particular solution for large positive values of x.

3 a Find the value of k such that kx^2e^{-2x} is a particular integral of the differential equation

$$\dfrac{d^2y}{dx^2} + 4\dfrac{dy}{dx} + 4y = e^{-2x}$$

b Hence find the general solution of this equation.

The equation of a curve C satisfies this differential equation.
The curve crosses the y-axis at $y = 8$ and the x-axis at $x = 4$.

c i Show that the equation of C can be expressed as $y = \dfrac{1}{2}(x - 4)^2 e^{-2x}$

ii Sketch the graph of C.

4 a Find the general solution of the differential equation

$$\dfrac{d^2y}{d\theta^2} + 4\dfrac{dy}{d\theta} + 3y = 6\cos\theta - 12\sin\theta$$

b Find the particular solution $y(\theta)$ for which $y = 6$ and

$\dfrac{dy}{d\theta} = -5$ when $\theta = 0$.

c Show that $-3 < y(\theta) < 6$ for all $\theta > 0$.

5 By finding an appropriate particular integral, find the general solution of the differential equation

$$\frac{d^2x}{dt^2} + 2\frac{dx}{dt} + kx = 3t - 4$$

for constant k, when

a $k = -3$

b $k = 0$.

6 The equation of a curve C satisfies the differential equation

$$2x^2\frac{d^2y}{dx^2} - x(5x + 4)\frac{dy}{dx} + (3x^2 + 5x + 4)y - 40x^3e^{-x} \qquad (1)$$

a Show that the substitution $y = ux$ transforms equation (1) into the differential equation

$$2\frac{d^2u}{dx^2} - 5\frac{du}{dx} + 3u = 40\,e^{-x} \qquad (2)$$

b Find the general solution of equation (2).

The curve passes through the point $(\ln 4, \ln 16)$.
The gradient of the curve at this point is $\ln 4$.

c Show that the equation of C is given by

$$y = \frac{1}{4}x\left(e^{\frac{3}{2}x} - e^x + 16e^{-x}\right)$$

7 a Show that the substitution $y = xv$ transforms the differential equation

$$\frac{1}{2}x^2\frac{d^2y}{dx^2} - x\frac{dy}{dx} + (1 + 8x^2)y + 6x^3\sin 2x = 0 \qquad (1)$$

into the equation

$$\frac{d^2v}{dx^2} + 16v = -12\sin 2x \qquad (2)$$

b Find the general solution of equation (2).

A member C of the family of solution curves of equation (1) crosses the x-axis when $x = \frac{1}{6}\pi$ and $x = \frac{1}{2}\pi$.

c i Show that the equation of C may be expressed as
$y = x\sin 2x\,(2\cos 2x - 1)$

ii Hence find coordinates of all the points where the curve C crosses the x-axis for $0 \leqslant x \leqslant \pi$.

FP2

8 a If $v = xy$, where y is a function of x,

show that $\dfrac{dv}{dx} = x\dfrac{dy}{dx} + y$ and find an expression for $\dfrac{d^2v}{dx^2}$.

b Show that the substitution $v = xy$ transforms the differential equation

$$2x\frac{d^2y}{dx^2} + (4 + 5x)\frac{dy}{dx} + (5 + 3x)y = 6x + 13 \qquad (1)$$

into the equation

$$2\frac{d^2v}{dx^2} + 5\frac{dv}{dx} + 3v = 6x + 3 \qquad (2)$$

c Find the solution of equation (2).

d Hence write down

   i the general solution y of equation (1)

   ii the approximate value of y when x is large and positive.

9 a Find the general solution of the differential equation

$$3\frac{d^2y}{dt^2} + 2\frac{dy}{dt} - y = (t + 3)(t - 4)$$

b Find the particular solution of this differential equation

for which $y = 5$ and $\dfrac{dy}{dt} = -4$ when $t = 0$.

c For this particular solution, calculate the exact value of y
when $t = -3$.

10 Given that $2x \sin 3x$ is a particular integral of the differential equation

$$\frac{d^2y}{dx^2} + 9y = k\cos 3x$$

where k is a constant

a show that $k = 12$

b find the particular solution of this differential equation for
which $y = 0$ and $\dfrac{dy}{dx} = 6$ at $x = 0$.

The equation of a curve C is given by this particular solution.

c Find the exact coordinates of the points where C crosses the

x-axis in the interval $-\dfrac{1}{3}\pi \leqslant x \leqslant \dfrac{1}{3}\pi$.

11 a Verify that $y = 2xe^{-x}$ is a particular integral of the differential equation

$$2\frac{d^2y}{dx^2} + 3\frac{dy}{dx} + y = -2e^{-x}$$

b Find the particular solution of this differential equation for which $y = -1$ when $x = 0$ and for which $y = \ln 2$ when $x = \ln 4$.

c Hence find the exact value of y when $x = 1$.

12 a Show that the transformation $y = xv$ transforms the equation

$$x^2\frac{d^2y}{dx^2} - 2x\frac{dy}{dx} + 2(1 + 8x^2)y = 8x^5 + x^3 \qquad (1)$$

into the equation

$$\frac{d^2v}{dx^2} + 16v = 8x^2 + 1 \qquad (2)$$

b Solve equation (2) to find v as a function of x.

c Hence state the general solution of equation (1).

Summary

- The general solution of the differential equation $a\dfrac{d^2y}{dx^2} + b\dfrac{dy}{dx} + cy = 0$

  has two arbitrary constants and is called the complementary function (CF).

  5.1
- You can find the CF by solving the auxiliary equation

  $am^2 + bm + c = 0$ and using the nature of its roots α and β.

  5.1
- Any function y which satisfies the differential equation

  $a\dfrac{d^2y}{dx^2} + b\dfrac{dy}{dx} + cy = f(x)$ is a particular integral (PI) of the equation.

  5.2
- The general solution of $a\dfrac{d^2y}{dx^2} + b\dfrac{dy}{dx} + cy = f(x)$ is given by

  $y = CF + PI$, where the PI depends on $f(x)$.

  5.2
- You can use boundary conditions to find an exact solution.

  5.3
- You can use a substitution to simplify a second-order differential equation.

  5.4

Links

The motion of a mass m on a spring at a time t can be described using Newton's second law as a second order differential equation

$$mg - kx = m\dfrac{d^2x}{dt^2}$$

where k is the spring constant.

This type of motion is known as simple harmonic motion and occurs in many systems in the real world.

FP2

6

Maclaurin and Taylor series

This chapter will show you how to

○ express a function as a power series

○ express the solution of a differential equation as a power series.

Refer to **C3** and **C4**.

See Sections 0.1 and 0.5 ⊙.

Before you start

You should know how to:

1 Find the first and second derivatives of a function.

e.g. If $f(x) = xe^x$ find $f'(x)$ and $f''(x)$

Use the product rule on xe^x:

$f'(x) = x(e^x) + e^x(1)$

Factorise: $= (x + 1)e^x$

Use the product rule again:

$f''(x) = (x + 1)(e^x) + e^x(1)$

$= (x + 2)(e^x)$

Hence $f'(x) = (x + 1)e^x$, $f''(x) = (x + 2)e^x$

2 Expand a function using the binomial theorem.

e.g. express $(1 + 2x)^{-1}$ in ascending powers of x up to and including the term in x^2.

Apply the binomial theorem:

$(1 + x)^n = 1 + nx + \dfrac{n(n-1)}{2!}x^2 + \cdots$

So $(1 + 2x)^{-1} = 1 + (-1)(2x) + \dfrac{(-1)(-2)}{2!}(2x)^2 + \cdots$

i.e. $(1 + 2x)^{-1} = 1 - 2x + 4x^2 + \cdots$

3 Use implicit differentiation.

e.g. Find an expression for $\dfrac{d}{dx}(y^4)$.

Use the chain rule:

$\dfrac{d}{dx}(y^4) = \dfrac{d}{dy}(y^4) \times \dfrac{dy}{dx}$

$= 4y^3 \dfrac{dy}{dx}$

Hence $\dfrac{d}{dx}(y^4) = 4y^3 \dfrac{dy}{dx}$

Check in:

1 Find the first and second derivatives of these functions. Simplify answers where appropriate.

a $f(x) = \sin^2 x$

b $f(x) = \ln(1 + 2x)$

c $f(x) = x^2 \ln x$

d $f(x) = \ln(\cos x)$

2 a Expand these functions up to and including the term in x^3.

i $(1 + 3x)^{-2}$ **ii** $\sqrt{1 - x}$ **iii** $\dfrac{x^2 - 1}{1 + 4x}$

b i Show that, for small values of x,

$$\sqrt{\left(1 + \frac{1}{2}x\right)} \approx 1 + \frac{1}{4}x - \frac{1}{32}x^2$$

ii By substituting $x = \frac{2}{3}$ into this expansion obtain an estimate for $\sqrt{3}$. Give your answer to 2 decimal places.

3 Use implicit differentiation to find an expression for these derivatives.

a $\dfrac{d}{dx}(y^5)$

b $\dfrac{d}{dx}(x^2 y^2)$

c $\dfrac{d}{dx}\left(x^3 \dfrac{dy}{dx}\right)$

d $\dfrac{d}{dx}\left(y^2 \dfrac{dy}{dx}\right)$

FP2

You can repeatedly differentiate a function to find its first, second and higher derivatives.

For the function $f(x)$, $f^{(n)}(x)$, where $n \in \mathbb{N}$, is the nth derivative of $f(x)$.

In this section, all derivatives are with respect to x.

If $y = f(x)$, then the nth derivative of y is written as $\dfrac{d^n y}{dx^n}$.

$f^{(0)}(x) = f(x)$

EXAMPLE 1

Find the first, second, and third derivatives of the function
$$f(x) = \sin(2x + 1)$$

$f(x) = \sin(2x + 1)$
The first derivative of $f(x)$, $f^{(1)}(x) = 2\cos(2x + 1)$
The second derivative of $f(x)$, $f^{(2)}(x) = -4\sin(2x + 1)$
The third derivative of $f(x)$, $f^{(3)}(x) = -8\cos(2x + 1)$

Use the chain rule.
See **C3** for revision.

EXAMPLE 2

If $y = e^{\sin x}$

a find $\dfrac{d^2 y}{dx^2}$ **b** evaluate $\dfrac{d^2 y}{dx^2}$ when $x = \pi$.

a Find the first and second derivatives of y:

$$y = e^{\sin x} \quad \text{so} \quad \frac{dy}{dx} = e^{\sin x}\cos x$$

$$\text{Hence} \quad \frac{d^2 y}{dx^2} = \frac{d}{dx}(e^{\sin x}\cos x)$$

Use the product and chain rules:

$$\frac{d}{dx}(e^{\sin x}\cos x) = e^{\sin x}(-\sin x) + \cos x(e^{\sin x}\cos x)$$

Factorise: $\dfrac{d^2 y}{dx^2} = e^{\sin x}(\cos^2 x - \sin x)$

b Substitute the given value of x into $\dfrac{d^2 y}{dx^2}$:

When $x = \pi$, $\dfrac{d^2 y}{dx^2} = e^{\sin(\pi)}(\cos^2(\pi) - \sin(\pi))$

$$= e^0((-1)^2 - 0) = 1$$

If $y = f(x)$ and x_0 is a particular value of x then

$\left(\dfrac{d^n y}{dx^n}\right)_{x_0}$ is the value of $\dfrac{d^n y}{dx^n}$ at $x = x_0$.

This is also written as $f^{(n)}(x_0)$.

In Example 2, $\left(\dfrac{d^2 y}{dx^2}\right)_\pi = 1$

Exercise 6.1

1 Find the first, second and third derivatives of these functions.

    a $f(x) = e^{2x+1}$ **b** $f(x) = \cos(1 - 3x)$

    c $f(x) = \ln(x + 3)$ **d** $f(x) = \sqrt{2x + 1}$

    e $f(x) = xe^{2x}$ **f** $f(x) = x^2 \ln x$

2 Find $\left(\dfrac{d^2 y}{dx^2}\right)_{x_0}$ for each of these.

    a $y = 2e^{-x}$, $x_0 = 0$ **b** $y = \sqrt[3]{3x + 2}$, $x_0 = -1$

    c $y = \ln(\sin x)$, $x_0 = \dfrac{1}{2}\pi$

3 Find the exact value of $f^{(2)}(x_0)$ for each of these.

    a $f(x) = \ln(x^2 + 1)$, $x_0 = 0$ **b** $f(x) = \sin(x^2)$, $x_0 = \sqrt{\pi}$

    c $f(x) = e^x \ln x$, $x_0 = 1$ **d** $f(x) = e^{(x^2)}$, $x_0 = -1$

4 $y = \sqrt{4x + k}$ where k is a constant. It is given that $\left(\dfrac{dy}{dx}\right)_1 = \dfrac{2}{3}$

    a Show that $k = 5$.

    b Find the value of $\left(\dfrac{d^2 y}{dx^2}\right)_1$ and the value of $\left(\dfrac{d^3 y}{dx^3}\right)_{-1}$

5 For the function $f(x) = 1 + 2x + 3x^2 + 4x^3$

    a show that $f^{(1)}(0) = 2$

    b verify that $f(x) \equiv f^{(0)}(0) + f^{(1)}(0)\dfrac{x}{1!} + f^{(2)}(0)\dfrac{x^2}{2!} + f^{(3)}(0)\dfrac{x^3}{3!}$

6 Given that $y = 2^x$

    a prove by induction that

    $\dfrac{d^n y}{dx^n} = (\ln 2)^n 2^x$ for all $n \in \mathbb{N}$.

    b Hence find the value of the positive integer N such that

    $\left(\dfrac{d^N y}{dx^N}\right)_2 = (\ln 4)^2 \ln 2$

7 If $f(x) = \ln(1 - x)$

    a prove by induction that $f^{(n)}(x) = -(n - 1)!(1 - x)^{-n}$ for all $n \in \mathbb{N}$

    b simplify $\dfrac{f^{(n+1)}\left(\dfrac{1}{2}\right)}{f^{(n)}\left(\dfrac{1}{2}\right)}$ for $n \in \mathbb{N}$.

Provided $f^{(n)}(0)$ can be evaluated for all $n \geqslant 0$, you can express the function $f(x)$ as a power series.

> A power series is a series of ascending powers of x.

This series is known as the **Maclaurin expansion** of $f(x)$.

> The Maclaurin expansion of $f(x)$ is the infinite series
> $$f(x) = f^{(0)}(0) + \frac{x}{1!}f^{(1)}(0) + \frac{x^2}{2!}f^{(2)}(0) + \frac{x^3}{3!}f^{(3)}(0) + \cdots$$
> The expansion might be valid only for certain values of x.

> This result is in the **FP2** section of the formula book.
>
> $f^{(n)}(0) = n$th derivative of $f(x)$ evaluated at $x = 0$.

EXAMPLE 1

Find the first four terms in ascending powers of x of the Maclaurin expansion of the function $f(x) = e^x$

Find the values of $f^{(n)}(0)$ for $n = 0, 1, 2, 3$:

$$f(x) = e^x \quad \text{so} \quad f^{(0)}(0) = e^0 = 1$$
$$f(x) = e^x \quad \text{so} \quad f^{(1)}(x) = e^x \text{ and hence } f^{(1)}(0) = 1$$
$$f^{(1)}(x) = e^x \quad \text{so} \quad f^{(2)}(x) = e^x \text{ and hence } f^{(2)}(0) = 1$$

Similarly $f^{(3)}(0) = 1$

> $f^{(0)}(x)$ means $f(x)$
>
> $\frac{d}{dx}(e^x) = e^x$
>
> In fact $f^{(n)}(0) = 1$ for all $n \geqslant 0$.

Use the general Maclaurin expansion:

$$f(x) = f^{(0)}(0) + \frac{x}{1!}f^{(1)}(0) + \frac{x^2}{1!}f^{(2)}(0) + \frac{x^3}{3!}f^{(3)}(0) + \cdots$$

$$= 1 + \frac{x}{1!} \times 1 + \frac{x^2}{2!} \times 1 + \frac{x^3}{3!} \times 1 + \cdots$$

i.e. $e^x = 1 + x + \frac{1}{2}x^2 + \frac{1}{6}x^3 + \cdots$

Some important functions and their Maclaurin expansions are:

$$e^x = 1 + x + \frac{x^2}{2!} + \frac{x^3}{3!} + \cdots + \frac{x^r}{r!} + \cdots \qquad \text{(valid for all } x \in \mathbb{R})$$

$$\ln(1 + x) = x - \frac{x^2}{2} + \frac{x^3}{3} + \cdots + (-1)^{r+1}\frac{x^r}{r} + \cdots \qquad \text{(valid only for } -1 < x \leqslant 1)$$

$$\sin x = x - \frac{x^3}{3!} + \frac{x^5}{5!} + \cdots + (-1)^r \frac{x^{2r+1}}{(2r+1)!} + \cdots \qquad \text{(valid for all } x \in \mathbb{R})$$

$$\cos x = 1 - \frac{x^2}{2!} + \frac{x^4}{4!} + \cdots + (-1)^r \frac{x^{2r}}{(2r)!} + \cdots \qquad \text{(valid for all } x \in \mathbb{R})$$

$$(1 + x)^n = 1 + nx + \frac{n(n-1)x^2}{2!} + \cdots + n(n-1) \cdots \frac{(n-r+1)x^2}{r!} + \cdots$$

$$(n \in \mathbb{R}) \quad \text{(valid only for } -1 < x < 1)$$

> These expansions are in the **FP2** section of the formula book. However, in the exam you may be asked to derive these series from first principles.
>
> This is the binomial theorem. See **C4** for revision.
>
> This result is given in the **C2** section of the formula book.

FP2

You can find approximate values of a function by using the first few terms of its Maclaurin expansion.

EXAMPLE 2

Use the Maclaurin expansion of e^x in ascending powers of x up to and including the term in x^3 to estimate the value of $\sqrt{e}$.

$e^x = 1 + x + \dfrac{x^2}{2!} + \dfrac{x^3}{3!} + \cdots + \dfrac{x^r}{r!} + \cdots$

See Example 1.

For small values of x the value of x^r becomes less significant as r increases.

Hence $e^x \approx 1 + x + \dfrac{x^2}{2!} + \dfrac{x^3}{3!}$ for small values of x.

This is a good approximation only for values of x close to 0.

Replace x with 0.5 in the approximation $e^x \approx 1 + x + \dfrac{x^2}{2!} + \dfrac{x^3}{3!}$:

$\sqrt{e} = e^{0.5}$

$e^{0.5} \approx 1 + (0.5) + \dfrac{(0.5)^2}{2!} + \dfrac{(0.5)^3}{3!}$

$= 1.645833\ldots$

Hence $\sqrt{e} \approx 1.645833\ldots$

The terms in x^4 and higher powers of x have been ignored.

Check this approximation with the value given by your calculator.

You can use standard expansions efficiently to find a Maclaurin expansion of more complicated functions.

FP2

EXAMPLE 3

Use standard expansions to find the first three non-zero terms in the Maclaurin expansions, in ascending powers of x, of these functions.
In each case state the range of values of x for which the full expansion is valid.
a $\cos 2x$
b $e^x - \cos x$
c $(2x + 3)\ln(1 + x)$

a The standard expansion for $\cos x$ is

$\cos x = 1 - \dfrac{x^2}{2!} + \dfrac{x^4}{4!} + \cdots$

Replace x with $2x$ to find the expansion of $\cos(2x)$:

$\cos(2x) = 1 - \dfrac{(2x)^2}{2!} + \dfrac{(2x)^4}{4!} + \cdots$

so $\cos 2x = 1 - 2x^2 + \dfrac{2}{3}x^4 + \cdots$

Use brackets to help you expand correctly.

The expansion is valid for all $x \in \mathbb{R}$ since

$\cos x = 1 - \dfrac{x^2}{2!} + \dfrac{x^4}{4!} + \cdots$ for all $x \in \mathbb{R}$

Example 3 is continued on the next page.

EXAMPLE 3 (CONT.)

b $e^x = 1 + x + \dfrac{x^2}{2!} + \dfrac{x^3}{3!} + \dfrac{x^4}{4!} + \cdots$ and $\cos x = 1 - \dfrac{x^2}{2!} + \dfrac{x^4}{4!} + \cdots$

Subtract the terms in the expansion of $\cos x$ from those in the expansion of e^x to obtain the expansion of $e^x - \cos x$:

$$e^x - \cos x = \left(1 + x + \tfrac{1}{2}x^2 + \tfrac{1}{6}x^3 + \tfrac{1}{24}x^4 + \cdots\right) - \left(1 - \tfrac{1}{2}x^2 + \tfrac{1}{24}x^4 + \cdots\right)$$

$$= 1 + x + \tfrac{1}{2}x^2 + \tfrac{1}{6}x^3 + \tfrac{1}{24}x^4 + \cdots - 1 + \tfrac{1}{2}x^2 - \tfrac{1}{24}x^4 + \cdots$$

> You may need to include more terms in each series than are required in the final answer.
>
> Take care with the negative signs.

Collect like terms and simplify coefficients:

$$e^x - \cos x = x + x^2 + \tfrac{1}{6}x^3 + \cdots$$

> These are the first three non-zero terms in the required expansion.

This expansion is valid for all $x \in \mathbb{R}$ because the expansions for e^x and $\cos x$ are themselves valid for all $x \in \mathbb{R}$.

c $\ln(1 + x) = x - \dfrac{x^2}{2} + \dfrac{x^3}{3} - \dfrac{x^4}{4} + \cdots$ (for $-1 < x \leqslant 1$)

Multiply the series for $\ln(1 + x)$ by each term in the bracket $(2x + 3)$ to obtain the expansion of $(2x + 3)\ln(1 + x)$:

$$(2x + 3)\ln(1 + x) = (2x + 3)\left(x - \tfrac{1}{2}x^2 + \tfrac{1}{3}x^3 - \tfrac{1}{4}x^4 + \cdots\right)$$

$$= 2x\left(x - \tfrac{1}{2}x^2 + \tfrac{1}{3}x^3 - \tfrac{1}{4}x^4 + \cdots\right) + 3\left(x - \tfrac{1}{2}x^2 + \tfrac{1}{3}x^3 - \tfrac{1}{4}x^4 + \cdots\right)$$

$$= \left(2x^2 - x^3 + \tfrac{2}{3}x^4 - \tfrac{1}{2}x^5 + \cdots\right) + \left(3x - \tfrac{3}{2}x^2 + x^3 - \tfrac{3}{4}x^4 + \cdots\right)$$

$$= 3x + \left(2 - \tfrac{3}{2}\right)x^2 + (-1 + 1)x^3 + \left(\tfrac{2}{3} - \tfrac{3}{4}\right)x^4 + \cdots$$

> Collect like terms.

$$= 3x + \tfrac{1}{2}x^2 - \tfrac{1}{12}x^4 + \cdots$$

> The coefficient of x^3 is zero in this expansion.

Hence $(2x + 3)\ln(1 + x) = 3x + \tfrac{1}{2}x^2 - \tfrac{1}{12}x^4 + \cdots$

This expansion is valid for $-1 < x \leqslant 1$ since this is the restriction on x in the expansion of $\ln(1 + x)$.

Exercise 6.2

1 Use differentiation to find the Maclaurin expansion of these functions in ascending powers of x up to and including the term in x^3.

> In questions **1** and **2** you must find each expansion from first principles.

  a e^x **b** $\sin x$ **c** $\ln(1 + x)$ **d** $(1 + x)^{\frac{1}{2}}$

2 Use differentiation to find the Maclaurin expansion of these functions in ascending powers of x up to and including the term in x^3. Leave coefficients in exact form where appropriate.

  a $\sin\left(x + \tfrac{1}{6}\pi\right)$ **b** $\cos\left(2x + \tfrac{1}{3}\pi\right)$ **c** $\ln(x + e)$

3 Find the first three non-zero terms of the Maclaurin expansion of these functions, in ascending powers of x. In each case, state the range of values of x for which the full expansion is valid.

For question 3 onwards, you may, unless told otherwise, use standard Maclaurin expansions.

a e^{-2x} b $\cos 4x$

c $\ln(1 + x^2)$ d $\sqrt{e^{3x}}$

4 a Find the Maclaurin expansion of $\sin 3x$ in ascending powers of x up to and including the term in x^5.

    b Hence, by substituting $x = \frac{1}{3}$ into this expansion, show that $\sin 1^c \approx \frac{101}{120}$

1^c means 1 radian.

5 Find the Maclaurin expansion of these functions in ascending powers of x up to and including the term in x^3. In each case, state the range of values of x for which the full expansion is valid.

a $\sin 2x + \cos 2x$ b $(x^2 + 1)e^{2x}$

c $(1 + x)^2 \ln(1 + 2x)$ d $\dfrac{1}{(1 + x)e^x}$

6 a Use a suitable Maclaurin expansion to show that
$$e^{-x} \approx 1 - x + \frac{1}{2}x^2 - \frac{1}{6}x^3 \quad \text{for small values of } x.$$

    b Use the result from part a to find an approximate value for $\dfrac{1}{\sqrt[3]{e}}$. Give your answer to 5 decimal places.

7 a Find, by differentiation, the first two non-zero terms in ascending powers of x of the Maclaurin expansion of $(2x + 1)\ln(2x + 1)$.

It is given that, in the interval $(-0.5, 0)$, the equation $(2x + 1)\ln(2x + 1) + 10x^2 = 0$ has exactly one root, α.

    b Use the result of part a to estimate α.

8 Use standard Maclaurin expansions of these functions to find the expansion, in ascending powers of x up to and including the term in x^3.

a $\ln(1 + x)\sin 2x$ b $e^{2x}\cos 4x$ c $\sqrt{e^x}\ln(1 - x)$

FP2

9 a Show that $\sqrt{2}e^x\sin\left(x - \frac{1}{4}\pi\right) \equiv e^x(\sin x - \cos x)$

b Hence, or otherwise, find, in ascending powers of x, the Maclaurin expansion of

$$\sqrt{2}\,e^x\sin\left(x - \frac{1}{4}\pi\right)$$

up to and including the term in x^3.

10 a Find, from first principles, the Macluarin expansion of e^{ax+b}, where a and b are constants, in ascending powers of x up to and including the term in x^2. Give coefficients in terms of a and b.

It is given that $e^{ax+b} \equiv k + 6x + 9x^2 + \cdots$, where k is a constant

b Find the values of the constants a, k and b. Give the answer for b in exact form.

11 a Use a suitable Maclaurin expansion to show that

$$\ln\left(\frac{1}{1-x}\right) = x + \frac{1}{2}x^2 + \frac{1}{3}x^3 + \cdots$$

and state the range of values of x for which the full expansion is valid.

b Assuming the pattern in this expansion continues, find the value of the infinite series $\displaystyle\sum_{r=1}^{\infty} \frac{1}{r\,2^r}$

12 a Use the binomial theorem to find the first two terms in ascending powers of x of the expansion of $(1 - kx^2)^{-1}$ where k is a positive constant.

b By considering a suitable Maclaurin expansion, deduce that
$$\sec 2x \approx 1 + 2x^2 \quad \text{for small values of } x.$$

It is given that in the interval $(0, 0.5)$, the equation $4x - \sec 2x = 0$ has exactly one root α.

c Use the result of part b to find an approximate value of α. Show, by means of a change of sign, that your answer is accurate to at least 1 decimal place.

13 $f(x) = \ln(1 + \sin x)$, where $|\sin x| < 1$

    a Show that $f^2(x) = -\dfrac{1}{(1 + \sin x)}$

    b Find the Maclaurin series for $f(x)$ in ascending powers of x up to and including the term in x^3.

    c Hence write down the first three non-zero terms in ascending powers of x in the Maclaurin expansion of $\ln(1 - \sin x)$.

    $\sin(-x) \equiv -\sin x$

    d Given that $\cos x > 0$, deduce that the Maclaurin expansion of $\ln(\cos x)$ in ascending powers of x is given by

        $\ln(\cos x) - \dfrac{1}{?}x^2 + \cdots$

    e Verify the result of part **d** from first principles.

14 **a** If $x = \tan y$, show that $\dfrac{dy}{dx} = \dfrac{1}{1 + x^2}$

    b Use differentiation to show that

        $\arctan x \approx x - \dfrac{1}{3}x^3$ for small values of x.

It can be shown that the full Maclaurin expansion of $\arctan x$ in ascending powers of x is given by

    $\arctan x = \displaystyle\sum_{r=0}^{\infty} \dfrac{p^r}{2r + q} x^{2r+1}$, where p and q are constants.

    c Use the result from part **b** to find the value of p and q.

    Hence evaluate the infinite series

        $\dfrac{2}{3} - \dfrac{2}{5} + \dfrac{2}{7} - \dfrac{2}{9} + \cdots$

    giving your answer in terms of π.

15 **a** Write down the first six terms, in ascending powers of x, of the series expansion for e^x.

    b Assuming this expansion of e^x is valid when $x = i\theta$, for $\theta \in \mathbb{R}$ in radians, show that

        $e^{i\theta} \equiv \cos\theta + i\sin\theta$

FP2

For small values of x, you can approximate a function $f(x)$ by the first few terms of its Maclaurin expansion.

If, instead, you need to approximate $f(x)$ for values of x not close to 0, you can use a **Taylor expansion** for $f(x)$.

> If a is a constant then the Taylor expansion of $f(x)$ in ascending powers of $(x - a)$ is
>
> $$f(x) = f(a) + (x - a)f^{(1)}(a) + \frac{(x - a)^2}{2!}f^{(2)}(a) + \cdots + \frac{(x - a)^r}{r!}f^{(r)}(a) + \cdots$$

This result is in the **FP2** section of the formula book.

When $a = 0$, the Taylor and Maclaurin expansions of $f(x)$ are identical.

For $x \approx a$ the first few terms of this Taylor series give a good approximation to $f(x)$.

If $x \approx a$ then each term $(x - a)^r$ in the expansion will become less significant as r increases.

EXAMPLE 1

a Find the first two non-zero terms in the Taylor expansion of $\cos x$ in ascending powers of $(x - \pi)$.

b Hence obtain an estimate of $\cos 3^c$

a Define $f(x) = \cos x$

Compare $(x - \pi)$ with $(x - a)$.

Evaluate the function and its derivatives at $x = \pi$:

$f(\pi) = \cos \pi = -1$

$f^{(1)}(x) = -\sin x$ so $f^{(1)}(\pi) = -\sin \pi = 0$

$f^{(2)}(x) = -\cos x$ so $f^{(2)}(\pi) = -\cos \pi = 1$

The second term of the Taylor expansion is zero so continue the process.

Use the general Taylor expansion for a function:

$$f(x) = f(a) + (x - a)f^{(1)}(a) + \frac{(x - a)^2}{2!}f^{(2)}(a) + \cdots$$

$\cos x = (-1) + \frac{(x - \pi)^2}{2!}(1) + \cdots$

$\qquad = -1 + \frac{1}{2}(x - \pi)^2 + \cdots$

b Since $3 \approx \pi$, substitute $x = 3$ into the expression in part **a** to estimate $\cos 3$:

$\cos 3^c \approx -1 + \frac{1}{2}(3 - \pi)^2$

$\qquad\quad = -0.9899757602\ldots$

Check the accuracy of this approximation with the value given by your calculator.

FP2

Replacing x by the term $(x + a)$ in the Taylor expansion, where a is a constant, gives

$$f(x + a) = f(a) + xf^{(1)}(a) + \frac{x^2}{2!}f^{(2)}(a) + \cdots + \frac{x^r}{r!}f^{(r)}(a) + \cdots$$

This result is in the **FP2** section of the formula book.

For small values of x, the first few terms of this expansion give a good approximation for $f(x + a)$.

EXAMPLE 2

Express $\sin\left(x + \frac{1}{6}\pi\right)$ in ascending powers of x up to and including the term in x^2.

Compare $\sin\left(x + \frac{1}{6}\pi\right)$ with $f(x + a)$: $f(x) = \sin x, \quad a = \frac{1}{6}\pi$

Evaluate $f\left(\frac{1}{6}\pi\right), f^{(1)}\left(\frac{1}{6}\pi\right)$ and $f^{(2)}\left(\frac{1}{6}\pi\right)$:

$$f(x) = \sin x \quad \text{so} \quad f\left(\frac{1}{6}\pi\right) = \frac{1}{2}$$

$$f^{(1)}(x) = \cos x \quad \text{so} \quad f^{(1)}\left(\frac{1}{6}\pi\right) = \frac{\sqrt{3}}{2}$$

$$f^{(2)}(x) = -\sin x \quad \text{so} \quad f^{(2)}\left(\frac{1}{6}\pi\right) = -\frac{1}{2}$$

Use the expansion $f(x + a) = f(a) + xf^{(1)}(a) + \frac{x^2}{2!}f^{(2)}(a) + \cdots$:

$$\sin\left(x + \frac{1}{6}\pi\right) = \frac{1}{2} + x\frac{\sqrt{3}}{2} + \frac{x^2}{2!}\left(-\frac{1}{2}\right) + \cdots$$

$$= \frac{1}{2} + \frac{\sqrt{3}}{2}x - \frac{1}{4}x^2 + \cdots$$

This is the same answer as you would obtain by finding the Maclaurin expansion of $\sin\left(x + \frac{1}{6}\pi\right)$.
Check this for yourself.

Exercise 6.3

1 Find the Taylor expansion of these functions in ascending powers of $(x - a)$, for the given value of the constant a, up to and including the term in $(x - a)^3$.

a $\sin x, a = \pi$ b $\cos x, a = \frac{3}{2}\pi$ c $e^{2x}, a = \ln 2$

In all questions use, where appropriate, a suitable Taylor series to find the required expansion.

2 Given that $y = \ln x$,

a find $\dfrac{dy}{dx}, \dfrac{d^2y}{dx^2}$ and $\dfrac{d^3y}{dx^3}$

b find the Taylor expansion in ascending powers of $(x - 1)$ of $\ln x$ up to and including the term in $(x - 1)^3$

c use a suitable value of x in your answer to **b** to estimate
    i $\ln 1.5$ ii $\ln 6 - \ln 5$
    Give each answer to 3 decimal places.

3 a Show that $\sin 2x = p(x - \pi) + q(x - \pi)^3 + \cdots$ for constants p and q to be stated.

   b Hence, or otherwise, find the expansion of $\sin x \cos x$ in ascending powers of $(x - \pi)$, up to and including the term in $(x - \pi)^3$.

4 a Show that

$$\cos 2x \approx -1 + 2\left(x - \tfrac{1}{2}\pi\right)^2 - \tfrac{2}{3}\left(x - \tfrac{1}{2}\pi\right)^4 \quad \text{for values of } x \text{ close to } \tfrac{1}{2}\pi.$$

   b Hence, or otherwise, find the first three non-zero terms in ascending powers of $\left(x - \tfrac{1}{2}\pi\right)$ in the Taylor expansion of $\sin^2 x$.

5 a Show that $\cos x = -\left(x - \tfrac{1}{2}\pi\right) + \tfrac{1}{6}\left(x - \tfrac{1}{2}\pi\right)^3$, provided $\left(x - \tfrac{1}{2}\pi\right)$

is so small that $\left(x - \tfrac{1}{2}\pi\right)^4$ and higher powers can be neglected.

   b Hence estimate $\cos 86°$. Use the approximation $1^c \approx 57\tfrac{1}{3}^\circ$ and give your answer to 2 decimal places.

6 $f(x) = (\ln x)^2, \quad x > 0$

   a Show that $f'(1) = 0$

   b Given that x is sufficiently close to 1 for $(x - 1)^4$ and higher powers of $(x - 1)$ to be ignored, show that
$$f(x) \approx (x - 1)^2 - (x - 1)^3$$

   c Estimate the value of $\displaystyle\int_{0.95}^{1.05} (6 \ln x)^2 \, dx$

7 a Use an appropriate expansion to express $\cos\left(x + \tfrac{1}{3}\pi\right)$ in ascending powers of x up to and including the term in x^2.

                                              Use the result $f(x + a) = f(a) + x f^{(1)}(a) + \cdots$

   b Hence show that, for small values of x,
$$\cos x - \sqrt{3}\sin x \approx 1 - \sqrt{3}x - \tfrac{1}{2}x^2$$

8 $f(x) = \dfrac{1}{x^2}, \quad x \neq 0$

   a Find the value of the constant a for which
$$f(x + a) \equiv \frac{1}{x^2 + 4x + 4}, \quad x \neq -a$$

   b Hence, or otherwise, obtain the expansion of
$$\frac{x}{x^2 + 4x + 4}$$
in ascending powers of x up to and including the term in x^3.

9 **a** Show that $2^{\ln x} = x^{\ln 2}$ for all $x > 0$.

b Hence find the Taylor expansion of $2^{\ln x}$ in ascending powers of $(x-1)$ up to and including the term in $(x-1)^2$. Give coefficients in exact form.

10 The Taylor expansion of $\sin(kx)$, where $0 < k < 5$ is a constant, in ascending powers of $\left(x - \frac{1}{4}\pi\right)$ is given by

$$\sin(kx) = A\left(x - \frac{1}{4}\pi\right) + B\left(x - \frac{1}{4}\pi\right)^3 + \cdots$$

where A and B are constants.

a Show that $k = 4$.

b Find the value of A and the value of B.

c Find the next non-zero term in this expansion.

11 **a** Show that the Taylor expansion of $\tan x$, in ascending powers of $\left(x - \frac{1}{4}\pi\right)$ up to and including the term in $\left(x - \frac{1}{4}\pi\right)^2$, is given by

$$\tan x = 1 + 2\left(x - \frac{1}{4}\pi\right) + 2\left(x - \frac{1}{4}\pi\right)^2 + \cdots$$

b Use this answer to estimate the value of $\int_{0.7}^{0.8} \sqrt{2\tan x - 1}\, dx$
Give your answer to 3 decimal places.

12 The function $f(x)$ is such that the first and higher derivatives of $f(x)$ exist for all x. It is given that $g(x) = f(x + a)$ where a is a constant.

a Prove by induction that, for all $n \geqslant 1$,
$$g^{(n)}(x) \equiv f^{(n)}(x + a)$$

b Deduce that $g^{(n)}(0) = f^{(n)}(a)$ for all $n \geqslant 0$.

c Use the Maclaurin expansion of $g(x)$ to show that
$$f(x + a) = f(a) + xf^{(1)}(a) + \frac{x^2}{2!} f^{(2)}(a) + \cdots$$

d Deduce the Taylor expansion of $f(x)$:
$$f(x) = f(a) + (x - a)f^{(1)}(a) + \frac{(x - a)^2}{2!} f^{(2)}(a) + \cdots$$

You can express the solution of a differential equation as a Taylor expansion.

$$y = y_a + \left(\frac{dy}{dx}\right)_a (x-a) + \left(\frac{d^2y}{dx^2}\right)_a \frac{(x-a)^2}{2!} + \left(\frac{d^3y}{dx^3}\right)_a \frac{(x-a)^3}{3!} + \cdots$$

y_a is the value of y when $x = a$.

where y (and possibly other values) at $x = a$ are known.

EXAMPLE 1

$\frac{dy}{dx} = x + y^2$, where $y = 1$ when $x = 0$.

This is a non-linear differential equation.

a Express the solution of this differential equation as a series of ascending powers of x up to and including the term in x^3.

b Hence estimate the value of y when $x = 0.15$.

----

a The solution $y(x)$ has a Taylor expansion

$$y = y_0 + \left(\frac{dy}{dx}\right)_0 x + \left(\frac{d^2y}{dx^2}\right)_0 \frac{x^2}{2!} + \cdots$$

This is also the Maclaurin expansion of $y(x)$.
$a = 0$

where y_0 is the value of y when $x = 0$, $\left(\frac{dy}{dx}\right)_0$ is the value of $\frac{dy}{dx}$ when $x = 0$, and so on.

When $x = 0$, $y = 1$ so $y_0 = 1$.

Use the differential equation $\frac{dy}{dx} = x + y^2$ to find $\left(\frac{dy}{dx}\right)_0$:

$$\frac{dy}{dx} = x + y^2 \text{ so, when } x = 0, \left(\frac{dy}{dx}\right)_0 = 0 + y_0^2$$
$$= 0 + 1^2$$
$$= 1$$

Differentiate each side of the equation $\frac{dy}{dx} = x + y^2$ with respect to x to obtain an expression for $\frac{d^2y}{dx^2}$:

$$\frac{dy}{dx} = x + y^2 \quad \text{so} \quad \frac{d^2y}{dx^2} = 1 + 2y\frac{dy}{dx}$$

Use implicit differentiation:
$\frac{d}{dx}(y^2) = \frac{d}{dy}(y^2) \times \frac{dy}{dx} = 2y\frac{dy}{dx}$

Hence $\left(\frac{d^2y}{dx^2}\right)_0 = 1 + 2y_0\left(\frac{dy}{dx}\right)_0$

Refer to **C4**.

$$= 1 + 2(1)(1)$$
$$= 3$$

EXAMPLE 1 (CONT.)

Differentiate each side of the equation $\dfrac{d^2y}{dx^2} = 1 + 2y\dfrac{dy}{dx}$ with respect to x to find an expression for $\dfrac{d^3y}{dx^3}$:

$$\frac{d}{dx}\left(1 + 2y\frac{dy}{dx}\right) = 2y\frac{d}{dx}\left(\frac{dy}{dx}\right) + \frac{dy}{dx}\left(2\frac{dy}{dx}\right)$$

$$= 2y\frac{d^2y}{dx^2} + 2\left(\frac{dy}{dx}\right)^2$$

$$\text{so } \frac{d^3y}{dx^3} = 2y\frac{d^2y}{dx^2} + 2\left(\frac{dy}{dx}\right)^2$$

Hence
$$\left(\frac{d^3y}{dx^3}\right)_0 = 2y_0\left(\frac{d^2y}{dx^2}\right)_0 + 2\left(\frac{dy}{dx}\right)_0^2$$

$$= 2\,(1)(3) + 2(1)^2$$

$$= 8$$

List the values to be used in the series expansion for y:

$$y_0 = 1, \quad \left(\frac{dy}{dx}\right)_0 = 1, \quad \left(\frac{d^2y}{dx^2}\right)_0 = 3, \quad \left(\frac{d^3y}{dx^3}\right)_0 = 8$$

Substitute these values into the Taylor expansion of $y(x)$:

$$y = y_0 + \left(\frac{dy}{dx}\right)_0 x + \left(\frac{d^2y}{dx^2}\right)_0 \frac{x^2}{2!} + \left(\frac{d^3y}{dx^3}\right)_0 \frac{x^3}{3!} + \cdots$$

so $\quad y = 1 + (1)x + (3)\dfrac{x^2}{2!} + (8)\dfrac{x^3}{3!} + \cdots$

Hence the series solution for the differential equation is
$$y = 1 + x + \frac{3}{2}x^2 + \frac{4}{3}x^3 + \cdots$$

b Substitute $x = 0.15$ into this series to estimate y:

$$y = 1 + x + \frac{3}{2}x^2 + \frac{4}{3}x^3 + \cdots$$

$$\approx 1 + (0.15) + \frac{3}{2}(0.15)^2 + \frac{4}{3}(0.15)^3$$

$$= 1.18825$$

i.e. when $x = 0.15$, $y \approx 1.18825$

You may assume that $x = 0.15$ is sufficiently small for this calculation to give a good approximation for y.

You can use a Taylor expansion to express the solution of a differential equation as a series of ascending powers of $(x - a)$ where a is a constant.

EXAMPLE 2

Given that y satisfies the differential equation $\dfrac{d^2y}{dx^2} + x\dfrac{dy}{dx} = y$, where $y = 5$ and $\dfrac{dy}{dx} = 2$ when $x = 1$, find the solution for y as a series of ascending powers of $(x-1)$ up to and including the term in $(x-1)^3$.

The appropriate Taylor expansion for y is

$$y = y_1 + \left(\frac{dy}{dx}\right)_1 (x-1) + \left(\frac{d^2y}{dx^2}\right)_1 \frac{(x-1)^2}{2!} + \left(\frac{d^3y}{dx^3}\right)_1 \frac{(x-1)^3}{3!} + \cdots$$

$a = 1$

Use $\dfrac{d^2y}{dx^2} + x\dfrac{dy}{dx} = y$ and the given values $x = 1$, $y_1 = 5$, $\left(\dfrac{dy}{dx}\right)_1 = 2$ to

evaluate $\left(\dfrac{d^2y}{dx^2}\right)_1$:

$$\left(\frac{d^2y}{dx^2}\right)_1 + (1)\left(\frac{dy}{dx}\right)_1 = y_1$$

i.e. $\left(\dfrac{d^2y}{dx^2}\right)_1 + (1)(2) = 5$

So $\left(\dfrac{d^2y}{dx^2}\right)_1 = 3$

Differentiate $\dfrac{d^2y}{dx^2} + x\dfrac{dy}{dx} = y$ with respect to x to find an expression for $\dfrac{d^3y}{dx^3}$:

Use the product rule on $x\dfrac{dy}{dx}$.

$$\frac{d^3y}{dx^3} + \left(x\frac{d^2y}{dx^2} + \frac{dy}{dx}\right) = \frac{dy}{dx}$$

Use the values $x = 1$, $\left(\dfrac{dy}{dx}\right)_1 = 2$ and $\left(\dfrac{d^2y}{dx^2}\right)_1 = 3$ to find $\left(\dfrac{d^3y}{dx^3}\right)_1$:

$$\left(\frac{d^3y}{dx^3}\right)_1 + ((1)(3) + 2) = 2$$

$$\left(\frac{d^3y}{dx^3}\right)_1 = -3$$

The values to be used in the series expansion for y are

$$y_1 = 5, \quad \left(\frac{dy}{dx}\right)_1 = 2, \quad \left(\frac{d^2y}{dx^2}\right)_1 = 3, \quad \left(\frac{d^3y}{dx^3}\right)_1 = -3$$

The solution for y is given by

$$y = y_1 + \left(\frac{dy}{dx}\right)_1 (x-1) + \left(\frac{d^2y}{dx^2}\right)_1 \frac{(x-1)^2}{2!} + \left(\frac{d^3y}{dx^3}\right)_1 \frac{(x-1)^3}{3!} + \cdots$$

$$= 5 + (2)(x-1) + (3)\frac{(x-1)^2}{2!} + (-3)\frac{(x-1)^3}{3!} + \cdots$$

$$y = 5 + 2(x-1) + \frac{3}{2}(x-1)^2 - \frac{1}{2}(x-1)^3 + \cdots$$

Exercise 6.4

1 Find a series solution of these differential equations in ascending powers of x up to and including the term in x^3.

   a $\dfrac{dy}{dx} = x^2 - y$ given that $y = 1$ when $x = 0$.

   b $\dfrac{dy}{dx} = x^2 + y^2$ given that $y = -2$ when $x = 0$.

   c $\dfrac{dy}{dx} = x + y^3$ given that $y = -1$ when $x = 0$.

2 It is given that y satisfies the differential equation

$$(x+1)\dfrac{dy}{dx} = 1 + xy$$

where $y = \dfrac{1}{2}$ when $x = 0$.

   a Find a series solution of this differential equation in ascending powers of x up to and including the term in x^3.

   b Hence estimate the value of y at $x = 0.1$.

3 Find a series solution of these differential equations in ascending powers of $(x - 1)$ up to and including the term in $(x - 1)^3$.

   a $\dfrac{dy}{dx} = x^2 + 2xy$ given that $y = 1$ when $x = 1$.

   b $(2x + 1)\dfrac{dy}{dx} = 2x^2 + y^2$ given that $y = 2$ when $x = 1$.

4 The equation of a curve C satisfies the differential equation

$$(x^2 - 1)\dfrac{dy}{dx} = y + 6x^2$$

It is given that the curve passes through the point $(0, -2)$.

   a Find a series solution of this differential equation in ascending powers of x up to and including the term in x^3.

   b Hence estimate the value of y when $x = 0.2$.

   c Use the differential equation to estimate the gradient of the curve C at $x = 0.2$.
   Give your answer to 2 decimal places.

5 Find a series solution of these differential equations in ascending powers of x up to and including the term in x^3.

a $\dfrac{d^2y}{dx^2} = x + y^2$ given that $y = 1$ and $\dfrac{dy}{dx} = 1$ when $x = 0$.

b $\dfrac{d^2y}{dx^2} + x\dfrac{dy}{dx} = 2y$ given that $y = 2$ and $\dfrac{dy}{dx} = -1$ when $x = 0$.

c $\dfrac{d^2y}{dx^2} + y\dfrac{dy}{dx} = x^2 + 1$ given that $y = 0$ and $\dfrac{dy}{dx} = 2$ when $x = 0$.

6 For the differential equation $(x - 1)\dfrac{d^2y}{dx^2} - x\dfrac{dy}{dx} + y^2 = x^2$

it is given that $y = 3$ and $\dfrac{dy}{dx} = 1$ when $x = 0$.

a Show that $\left(\dfrac{d^2y}{dx^2}\right)_0 = 9$

b Find $\left(\dfrac{d^3y}{dx^3}\right)_0$

c Express y as a series of ascending powers of x up to and including the term in x^3.

d Estimate y when $x = 0.2$. Give your answer to 2 decimal places.

7 Find a series solution of these differential equations in ascending powers of t up to and including the term in t^3.

a $\dfrac{dx}{dt} = t + \cos x$ given that $x = \pi$ when $t = 0$.

b $\dfrac{d^2x}{dt^2} + t^2\dfrac{dx}{dt} + x = xt - 1$ given that $x = 3$ and $\dfrac{dx}{dt} = -2$ when $t = 0$.

8 The equation of a curve C satisfies the differential equation

$$x^2\dfrac{d^2y}{dx^2} + y\dfrac{dy}{dx} + x^2 = 2$$

The curve has a stationary point at $P(1, 1)$.

a Find $y_1^{(2)}$ and hence state the nature of the stationary point at P.

b Show that $\left(\dfrac{d^3y}{dx^3}\right)_1 = -5$

c Hence express y as a series of ascending powers of $(x - 1)$ up to and including the term in $(x - 1)^3$.

d Estimate, to 3 decimal places, the value of y when
   i $x = 0.9$ **ii** $x = 1.1$
and comment on whether these estimates are consistent with the nature of point P as stated in **a**.

9 Given that $\frac{dy}{dx} = 2xy$ where $y = 1$ when $x = 0$

    a show that the first three non-zero terms in a series solution of this differential equation in ascending powers of x is given by

        $$y = 1 + x^2 + \frac{1}{2}x^4 + \cdots$$

> Separate the variables.
> See **C4**.

    b verify the answer to part **a** by finding the particular solution of this differential equation.

10 Given that $\frac{d^2y}{dx^2} + y^2\frac{dy}{dx} = x$, where $y = 1$ and $\frac{dy}{dx} = -1$ when $x = 0$,

    a show that $\left(\frac{d^2y}{dx^2}\right)_0 = 1$

    b find $\left(\frac{d^3y}{dx^3}\right)_0$

    c find the series solution of the given differential equation in ascending powers of x up to and including the term in x^4.

11 The function $y(x)$ satisfies the differential equation

    $$(2x+1)\frac{d^2y}{dx^2} + (2x+1)\frac{dy}{dx} + 2y = 0, \quad \text{where } y = 1 \text{ and } \frac{dy}{dx} = 1 \text{ at } x = 0$$

    a Show that $\left(\frac{d^2y}{dx^2}\right)_0 = -3$ and find $\left(\frac{d^3y}{dx^3}\right)_0$

    b Hence find the series solution of this differential equation in ascending powers of x up to and including the term in x^3.

It is given that the solution to this differential equation is given by

    $$y = (kx + h)e^{-x}$$

for constants k and h.

    c Find the value of k and the value of h.

    d Hence, or otherwise, express the function $(y(x))^2$ as a series of ascending powers of x up to and including the term in x^3.

FP2

1 a Find the first three non-zero terms in ascending powers of x of the Maclaurin expansion of $\cos 2x$.

b Hence, by using a suitable trigonometric identity, show that, for small x,
$$3\cos^2 x \approx a + bx^2 + cx^4$$
where a, b and c are integers to be determined.

c Find approximations for two roots of the equation
$$12\cos^2 x + 8x^2 - 11 = 0$$

2 $f(x) = \arcsin x, \quad -1 < x < 1$

a Given that $f^{(1)}(x) = \dfrac{1}{\sqrt{1-x^2}}$, find an expression for $f^{(2)}(x)$ and $f^{(3)}(x)$.
You need not simplify your answers.

b Hence find the first two non-zero terms in ascending powers of x of the Maclaurin expansion of $f(x)$.

c Use this approximation to find an estimate for
$$\int_0^{\frac{\sqrt{3}}{2}} \arcsin x\,\mathrm{d}x$$
Give your answer as a fraction in its lowest terms.

3 a Write down the Maclaurin expansion in ascending powers of x up to and including the term in x^3 of
i e^x
ii $\sin x$

b Hence show that, provided x is so small that x^4 and higher powers of x can be ignored,
$$e^x \sin x \approx \tfrac{1}{3}x(3 + 3x + x^2)$$

It is given that in the interval $(-1,0)$ the equation $3e^x \sin x - 2x = 0$ has exactly one real root α.

c Use the result of part **b** to find an estimate for α and show that your answer is accurate to at least 2 decimal places.

4 $f(x) = \ln(\sin 2x), \quad 0 < x < \frac{1}{2}\pi$

   a Find the first non-zero term in ascending powers of $\left(x - \frac{1}{4}\pi\right)$ in the Taylor expansion of $f(x)$.

   b Hence, or otherwise, show that, for values of x close to $\frac{1}{4}\pi$,

$$\ln(\operatorname{cosec} x) + \ln(\sec x) \approx \ln 2 + 2\left(x - \frac{1}{4}\pi\right)^2$$

5 $f(x) = \ln(2 - x), \quad x < 2$

   a Prove by induction that $f^{(n)}(x) \equiv -(n-1)!(2-x)^{-n}$ for all $n \geqslant 1$.

   b Hence write down the coefficient of x^4 in the Maclaurin expansion of $f(x)$.

   c Find the Taylor expansion of $f(x)$ in ascending powers of $(x - 1)$ up to and including the term in $(x - 1)^3$.

6 The Taylor expansion of $\ln x$ in ascending powers of $(x - 1)$ is given by

$$\ln x = (x - 1) - \frac{1}{2}(x - 1)^2 + \cdots$$

   a Find, by differentiation, the first three terms in the Taylor expansion in ascending powers of $(x - 1)$ of $\ln(x + 1)$.

   b Hence, or otherwise, show that, for values of x close to 1,

$$\ln\left(1 + \frac{1}{x}\right) \approx \ln 2 - \frac{1}{2}(x - 1) + \frac{3}{8}(x - 1)^2$$

7 **a** Use differentiation to find the Taylor expansion in ascending powers of $(x - \pi)$ up to and including the term in $(x - \pi)^3$ of $\sin 3x$.

   b Hence show that, for values of x close to, but not equal to, π,

$$(x - \pi)\operatorname{cosec} 3x \approx -\frac{1}{3}$$

8 It is given that the coefficient of x^4 in the Maclaurin expansion of $(1 + kx^2)\cos x$, where k is a constant, is $-\frac{23}{24}$.

   a Show that $k = 2$.

   b Find the coefficient of x^2 in this expansion.

   c Find the possible values of $p \in \mathbb{R}$ such that the coefficient of x^2 in the Maclaurin expansion of $(1 + x)^2 \cos px$ is zero.

9 $(1+x)\dfrac{dy}{dx} = x + y^2$ where $y = 2$ at $x = 0$.

  a Find the value of $\dfrac{dy}{dx}$ when $x = 0$.

  b Show that $(1+x)\dfrac{d^2y}{dx^2} = 1 + (2y - 1)\dfrac{dy}{dx}$

  c Hence find $\left(\dfrac{d^2y}{dx^2}\right)_0$ and show that $\left(\dfrac{d^3y}{dx^3}\right)_0 = 58$

  d Find a series solution of this differential equation in ascending powers of x up to and including the term in x^3.

10 The equation of a curve C satisfies the differential equation

  $$(2x+1)\dfrac{d^2y}{dx^2} + (x-1)\dfrac{dy}{dx} = y^2$$

  The curve has a stationary point at the point $(0, 1)$.

  a Show that $\left(\dfrac{d^2y}{dx^2}\right)_0 = 1$

  b Find the value of $\left(\dfrac{d^3y}{dx^3}\right)_0$

  c Hence find a series solution of this differential equation in ascending powers of x up to and including the term in x^3.

11 The variable y satisfies the differential equation

  $$\dfrac{d^2y}{dx^2} + (x+y)\dfrac{dy}{dx} - y^2 = x \quad \text{where } y = 2 \text{ and } \dfrac{dy}{dx} = -1 \text{ when } x = 1.$$

  a Show that $\dfrac{d^2y}{dx^2} = 8$ when $x = 1$.

  b Find an expression for $\dfrac{d^3y}{dx^3}$ in terms of x, y, $\dfrac{dy}{dx}$ and $\dfrac{d^2y}{dx^2}$.

  c Show that $\dfrac{d^3y}{dx^3} = -27$ when $x = 1$.

  d Hence express y as a series, in ascending powers of $(x-1)$, up to and including the term in $(x-1)^3$.

12 A curve C has equation which satisfies the differential equation

$$\cos x \frac{dy}{dx} + \sin y = x^2 + 2x + 3$$

C passes through the point $(0,0)$ and the point $(0.2, k)$
where k is a constant.

a Find a series solution of this differential equation, in ascending powers of x, up to and including the term in x^2.

b Hence find an estimate for the value of k.

13 The variable x satisfies the differential equation

$$\frac{d^2x}{dt^2} + x\frac{dx}{dt} + t^2 = t + 1, \quad \text{where } x = 1 \text{ and } \frac{dx}{dt} = 3 \text{ at } t = 0.$$

a Find the value of $\frac{d^2x}{dt^2}$ at $t = 0$.

b Show that $\left(\dfrac{d^3x}{dt^3}\right)_0 = -6$

c Hence show that if t is sufficiently small for t^4 and higher powers to be ignored then $x = 1 + 3t - t^2 - t^3$

14 It is given that the variable y satisfies the differential equation

$$(1 + 2x)\frac{dy}{dx} - 2y = 2x + \ln(1 + 2x), \quad \text{where } y = 0 \text{ when } x = 0.$$

a Show that a series solution for y is given by
$$y = 2x^2 - 2x^3 + \cdots$$

It is given that this series solution is of the form $y = x\ln(1 + kx)$,
for k a constant.

b Find the value of k.

c Hence find the coefficient of x^9 in this series solution.

6

Exit →

Summary

Refer to

- The Maclaurin expansion of $f(x)$ is the infinite series

$$f(x) = f^{(0)}(0) + \frac{x}{1!}f^{(1)}(0) + \frac{x^2}{2!}f^{(2)}(0) + \frac{x^3}{3!}f^{(3)}(0) + \cdots$$

6.2

- The Taylor expansion of $f(x)$ in ascending powers of $(x - a)$, where a is a constant, is

$$f(x) = f(a) + (x - a)f^{(1)}(a) + \frac{(x - a)^2}{2!}f^{(2)}(a) + \cdots + \frac{(x - a)^r}{r!}f^{(r)}(a) + \cdots$$

6.3

- The Taylor expansion of $f(x + a)$, where a is a constant, in ascending powers of x, is

$$f(x + a) = f(a) + xf^{(1)}(a) + \frac{x^2}{2!}f^{(2)}(a) + \cdots + \frac{x^r}{r!}f^{(r)}(a) + \cdots$$

6.3

- The series solution y of a differential equation is given by the Taylor expansion

$$y = y_a + \left(\frac{dy}{dx}\right)_a (x - a) + \left(\frac{d^2y}{dx^2}\right)_a \frac{(x - a)^2}{2!} + \left(\frac{d^3y}{dx^3}\right)_a \frac{(x - a)^3}{3!} + \cdots$$

where y (and possibly other values) at $x = a$ are known.

6.4

Links

Taylor and Maclaurin series are useful tools in the study of differential equations and numerical analysis, and provide highly accurate approximate answers where no exact solution can be found.

Computers and calculators do not calculate exact answers for values such as $\sin\left(\frac{\pi}{3}\right)$ or e^3. Instead they use a Taylor or Maclaurin expansion to calculate a required answer to the degree of accuracy that can be shown on the screen.

FP2

7

Polar coordinates

This chapter will show you how to
- plot polar points and sketch equations given in polar form
- convert between cartesian and polar forms
- deal with certain types of tangents to polar curves
- find the points where two polar curves intersect
- calculate an area swept out by a polar curve.

For background knowledge
see Sections 0.4, 0.5 and 0.6.

Before you start

You should know how to:

1 Use trigonometry to find the area and side lengths of a triangle.

e.g. Find the area of this triangle.

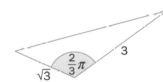

$$\text{Area} = \frac{1}{2}ab\sin C = \frac{1}{2} \times 3 \times \sqrt{3} \times \frac{\sqrt{3}}{2} = \frac{9}{4}$$

2 Differentiate and integrate trigonometric functions.

e.g. Find   a $\dfrac{dy}{d\theta}$ if $y = \sin^3 \theta$

b $\displaystyle\int 2\sin^2 \theta \, d\theta$

a $y = (\sin \theta)^3$ so $\dfrac{dy}{d\theta} = 3(\sin \theta)^2(\cos \theta)$

$= 3\sin^2\cos \theta$

b $\displaystyle\int 2\sin^2 \theta \, d\theta = \int (1 - \cos 2\theta) \, d\theta$

$= \theta - \dfrac{1}{2}\sin 2\theta + c$

3 Solve a trigonometric equation.

e.g. Solve $\sin 2\theta = \sin \theta$, $0 < \theta < \pi$

$\sin 2\theta = \sin \theta$ so $2\sin \theta \cos \theta - \sin \theta = 0$

i.e. $\sin \theta (2\cos \theta - 1) = 0$

For $0 < \theta < \pi$, $\sin \theta \neq 0$

and $\cos \theta = \dfrac{1}{2}$ only for $\theta = \dfrac{1}{3}\pi$

The solution is $\theta = \dfrac{1}{3}\pi$

Check in:

1 Find the area of the triangle and the exact length of the unknown side.

See **C2** for revision.

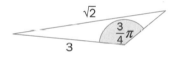

2 a Find, in terms of $\sin 4\theta$,

i $\dfrac{dx}{d\theta}$ for $x = \cos^2 2\theta$

ii $\dfrac{dy}{d\theta}$ for $y = \sin^2 \theta \cos^2 \theta$

See **C3** and **C4** for revision.

b Find

i $\displaystyle\int 4\cos^2 \theta \, d\theta$

ii $\displaystyle\int (2\sin \theta - 1)^2 \, d\theta$

3 Solve each equation exactly.

a $\cos 2\theta = \cos \theta$, $0 \leqslant \theta \leqslant \pi$

b $\sin 2\theta = 2\sin^2 \theta$, $-\pi \leqslant \theta \leqslant \pi$

c $\sin 2\theta - \cos 2\theta = 1$ for $-\dfrac{1}{2}\pi \leqslant \theta \leqslant \dfrac{1}{2}\pi$

You can describe the position of a point using polar coordinates.

The diagram shows a horizontal line *l* starting at the fixed point *O*.

O is called the pole. The line *l* is called the initial line.

The polar coordinates of *P* are (r, θ).

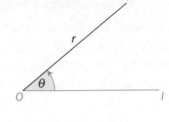

Length of
line *OP*

Angle (in radians) between
OP and the initial line

If point *P* has polar coordinates (r, θ) then:

- *r* is the length of *OP* and so $r \geqslant 0$
- θ is an angle in radians between the line *OP* and the initial line *l*
- angles measured in a clockwise direction are defined to be negative
- when $r = 0$, the point $P(0, \theta)$ is at the pole, for *any* value of θ.

The diagram shows points *A*, *B* and *C*.

Write down the polar coordinates of these points.

A has polar coordinates $\left(4, \frac{1}{3}\pi\right)$.

The angle between *l* and the line *OB* is $\frac{1}{3}\pi + \frac{4}{9}\pi = \frac{7}{9}\pi$

So *B* has polar coordinates $\left(4, \frac{7}{9}\pi\right)$.

Measured *clockwise*, the line *OC* has angle $-\frac{1}{2}\pi$ from the line *l*.

Hence *C* has polar coordinates $\left(3, -\frac{1}{2}\pi\right)$.

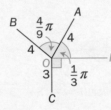

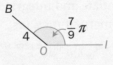

In polar coordinates, θ is not uniquely defined.

e.g. The polar coordinates of point *C* could also be written as $\left(3, \frac{3}{2}\pi\right)$

since $2\pi - \frac{1}{2}\pi = \frac{3}{2}\pi$

If the point *P* has polar coordinates (r, θ) then, unless told otherwise, assume that $-\pi < \theta \leqslant \pi$.

Compare with principal arguments
in complex numbers from **FP1**.

FP2

EXAMPLE 1

Exercise 7.1

1 Write down the polar coordinates of the points A, B, C and D.

Give answers in the form (r, θ) where $-\pi < \theta \leqslant \pi$ is exact.

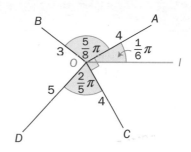

2 Express the polar coordinates of these points in the form (r, θ) where $-\pi < \theta \leqslant \pi$.

When not exact, give θ correct to 1 decimal place.

a $A\left(5, \frac{3}{2}\pi\right)$ b $B\left(3, \frac{7}{4}\pi\right)$ c $C\left(8, -\frac{4}{3}\pi\right)$ d $D(4, -\pi)$

e $E(1, 5^c)$ f $F(4, -4^c)$ g $G\left(\frac{1}{2}, \frac{17}{4}\pi\right)$ h $H\left(4, -\frac{13}{2}\pi\right)$

3 The diagram shows the points P and Q with polar coordinates

$P\left(3, \frac{5}{12}\pi\right)$ and $Q\left(4, \frac{3}{4}\pi\right)$.

a Show that angle $POQ = \frac{1}{3}\pi$.

b Hence find the exact length of PQ.

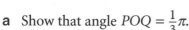

4 The diagram shows the points A, B and C with polar coordinates $A(2,0)$, $B(2, \pi)$ and $C(r, \theta)$ where $0 < \theta < \pi$.

A, B and C lie on a common circle centred at O.

a State the value of r.

b Given that $AC = 2\sqrt{3}$, show that $\theta = \frac{2}{3}\pi$.

c Without further calculation, find the length of BC.

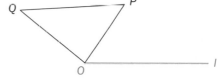

5 Points A and B have polar coordinates $A\left(5, \frac{1}{6}\pi\right)$, $B\left(6, -\frac{1}{2}\pi\right)$.

a Draw a diagram to show the positions of points A and B with respect to an initial line with pole O.

b Calculate the area of triangle AOB correct to 1 decimal place.

c Find the exact length of AB.

6 Points P and Q have polar coordinates $P\left(\sqrt{3}, \frac{1}{6}\pi\right)$ and $Q(r, \theta)$, where $-\pi < \theta \leqslant 0$.

It is given that the line PQ has length $2\sqrt{3}$ and is perpendicular to the initial line.

Find the polar coordinates of point Q.

You can sketch the graph of a polar equation by plotting points and joining them to make a smooth curve.

EXAMPLE 1

The polar curve C has equation $r = 1.5\theta + 1$ where $\theta \geqslant 0$.

a Copy and complete the table.
Give answers to 1 decimal place.

θ	0	$\frac{1}{4}\pi$	$\frac{2}{4}\pi$	$\frac{3}{4}\pi$	π
$r = 1.5\theta + 1$	1		3.4		5.7

By writing $\frac{2}{4}\pi$ rather than $\frac{1}{2}\pi$, you can see that θ is increasing in equal steps of $\frac{1}{4}\pi$.

b Hence sketch the curve C.

a Calculate the entries using the equation $r = 1.5\theta + 1$:

θ	0	$\frac{1}{4}\pi$	$\frac{1}{2}\pi$	$\frac{3}{4}\pi$	π
$r = 1.5\theta + 1$	1	2.2	3.4	4.5	5.7
Polar point	$P_1(1, 0)$	$P_2\left(2.2, \frac{1}{4}\pi\right)$	$P_3\left(3.4, \frac{1}{2}\pi\right)$	$P_4\left(4.5, \frac{3}{4}\pi\right)$	$P_5(5.7, \pi)$

b Because r is increasing with θ, the curve will spiral away from the pole.

Plot the five points and sketch the smooth curve which joins them:

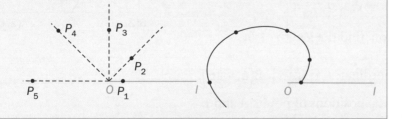

You do not need to use a scale to see the main features of the curve.

r cannot be negative. When choosing your own values of θ to sketch the polar curve $r = f(\theta)$, ensure that $r \geqslant 0$ over your chosen interval.

You can sketch certain standard curves by using the definition of a polar coordinate.

EXAMPLE 2

Sketch, on a separate diagram, the polar graph with

equation   **a** $r = 3$ **b** $\theta = \frac{1}{4}\pi$

a   Any polar point $P(3, \theta)$ on the curve must lie 3 units
from the pole.
The curve is therefore a circle, centre the pole,
radius 3 units.

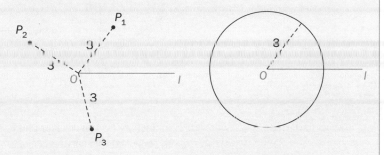

b   The polar point $P\left(r, \frac{1}{4}\pi\right)$ can be any distance from the
pole but must lie on a line which makes an angle of
$\frac{1}{4}\pi$ against l.

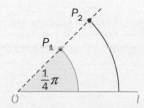 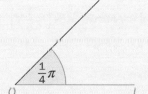

The equation $\theta = \frac{1}{4}\pi$ describes a **half-line** from O at

angle $\frac{1}{4}\pi$ against l.

The initial line l is the half-line $\theta = 0$.

You can use symmetry to complete the sketch of certain polar curves.

EXAMPLE 3

The diagram shows the polar curve given by the equation
    $r = 4\cos\theta$ for $0 \leqslant \theta \leqslant \frac{1}{2}\pi$

a   Sketch the graph of the polar curve C with equation
    $r = 4\cos\theta$ for $-\frac{1}{2}\pi \leqslant \theta \leqslant \frac{1}{2}\pi$

b   Find the exact polar coordinates of the point where the
half-line $\theta = \frac{1}{6}\pi$ intersects C.

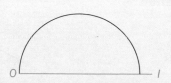

The solution to Example 3
is shown on the next page.

a The given diagram is the sketch of the curve C for $0 \leqslant \theta \leqslant \frac{1}{2}\pi$.

Since $4\cos(-\theta) \equiv 4\cos\theta$, the reflection in the initial line of any point $P(r, \theta)$ on the upper part of the curve will be a point $P'(r, -\theta)$ on the lower part of the curve, and vice versa.

> The graph of the equation $y = \cos\theta$ is symmetrical about the y-axis.
>
> See **C2** for revision.

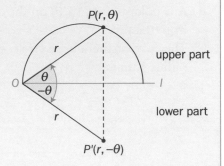

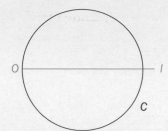

The curve C is therefore symmetrical about the initial line.

> In Section 7.3 you will learn how to show that this curve is a circle.

b The diagram shows the half-line $\theta = \frac{1}{6}\pi$ intersecting the curve C at point P.

The curve has equation $r = 4\cos\theta$.

Hence point P on this curve has polar coordinates $\left(r, \frac{1}{6}\pi\right)$ where $r = 4\cos\left(\frac{1}{6}\pi\right)$

$$= 4 \times \frac{1}{2}\sqrt{3} = 2\sqrt{3}$$

Hence the polar coordinates of P are $\left(2\sqrt{3}, \frac{1}{6}\pi\right)$.

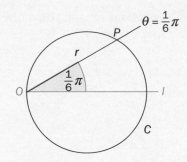

> You need to be familiar with the special angles. See **C2** for revision.

Standard polar curves

You should be familiar with the general shape of these standard polar curves, where $a > 0$ when stated.

Equation	Graph	Description/comment
$\theta = \alpha$	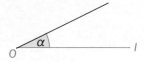	Half-line making an angle α with the intial line l
$r = k\sec(\alpha - \theta)$, $k > 0$	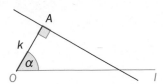	Straight line perpendicular to OA and passing through the polar point $A(k, \alpha)$

Equation	Graph	Description/comment
$r = a\theta$	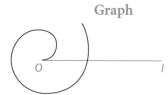	A spiral
$r = a$	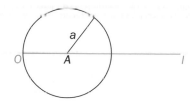	Circle, centre O and radius a
$r = 2a\cos\theta$	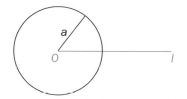	Circle, centre $A(a,0)$ and radius a
$r = 2a\sin\theta$	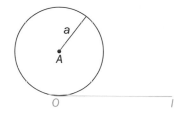	Circle, centre $A\left(a, \frac{1}{2}\pi\right)$ and radius a
$r = a(1 + \cos\theta)$	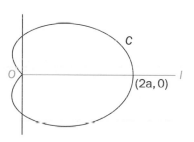	A cardioid The graph of $r = a(1 - \cos\theta)$ is a reflection of C in the half-lines $\theta = \pm\frac{1}{2}\pi$. The graph of $r = a(1 + \sin\theta)$ is an anti-clockwise rotation of C through $\frac{1}{2}\pi$ radians about O.
$r = 2a\cos 2\theta$		The graph of $r = 2a\sin 2\theta$ is an anti-clockwise rotation of this curve through $\frac{1}{4}\pi$ radians about O.
$r^2 = a^2\cos 2\theta$	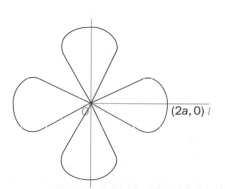	A lemniscate The graph of $r^2 = a^2\sin 2\theta$ is an anti-clockwise rotation of this curve through $\frac{1}{4}\pi$ radians about O.

FP2

Exercise 7.2

1 On separate diagrams, sketch these polar graphs.

   a $r = 5$ **b** $\theta = \frac{2}{3}\pi$ **c** $0.25r = 1$

   d $\theta = -\frac{2}{5}\pi$ **e** $\theta = \frac{11}{8}\pi$ **f** $\theta = 2.4^c$

2 **a** Sketch, on the same diagram, the graphs of the polar equations

      i $r = 6$ **ii** $\theta = \frac{1}{3}\pi$

   b Write down the polar coordinates of the point P where these two graphs intersect.

   c Find the polar coordinates of the two points on the circle $r = 6$ such that triangle OPQ is equilateral.

3 The polar curve C has equation $r = 3(1 + \cos\theta)$ for $-\pi \leqslant \theta \leqslant \pi$.

   a Copy and complete the table.
Give answers to 1 decimal place where appropriate.

θ	0	$\frac{1}{4}\pi$	$\frac{2}{4}\pi$	$\frac{3}{4}\pi$	π
$r = 3(1 + \cos\theta)$	6	5.1			

   b Sketch the graph of the curve $r = 3(1 + \cos\theta)$ for $0 \leqslant \theta \leqslant \pi$.

   c Using its symmetrical properties, sketch on a separate diagram the graph of C for $-\pi \leqslant \theta \leqslant \pi$.

4 A polar curve C has equation $r = 6 \sin\theta$ for $0 \leqslant \theta \leqslant \pi$.

   a Copy and complete the table.
Give answers to 1 decimal place where appropriate.

θ	0	$\frac{1}{8}\pi$	$\frac{2}{8}\pi$	$\frac{3}{8}\pi$	$\frac{4}{8}\pi$
$r = 6\sin\theta$	0	2.3			

   b Sketch the graph of $r = 6 \sin\theta$ for $0 \leqslant \theta \leqslant \frac{1}{2}\pi$.

   c Simplify $\sin(\pi - \theta)$ and hence show that the reflection in the half-line $\theta = \frac{1}{2}\pi$ of any point $P(r, \theta)$ on your sketch is a point on C.

   d Hence sketch on a separate diagram the graph of C.

5 The polar curve C is defined by the equation $r = \frac{10}{1 + \theta}$ for $0 \leqslant \theta \leqslant 2\pi$.

   a Copy and complete the table. Give answers to 1 decimal place.

θ	0	$\frac{1}{4}\pi$	$\frac{2}{4}\pi$	$\frac{3}{4}\pi$	π	$\frac{5}{4}\pi$	$\frac{6}{4}\pi$	$\frac{7}{4}\pi$	2π
$r = \frac{10}{1 + \theta}$	10	5.6		3.0			1.8		1.4

   b Sketch, on the same diagram, the graph of
      i C **ii** the half-line $\theta = 1.5^c$

   c Find the coordinates of the point where these graphs intersect.

6 A polar equation is given by $r = \sqrt{2}\sec\left(\frac{1}{4}\pi - \theta\right)$

a Copy and complete the table.
When not exact, give r to
1 decimal place.

θ	0	$\frac{1}{4}\pi$	$\frac{1}{2}\pi$
$r = \sqrt{2}\sec\left(\dfrac{1}{4}\pi - \theta\right)$			

b On the same diagram, plot the polar points P, Q and R

corresponding to the values $\theta = 0$, $\frac{1}{4}\pi$ and $\frac{1}{2}\pi$ respectively.

c Using the *exact* polar coordinates of these points, show that $PR - PQ + QR$
and hence state the geometrical relationship between P, Q and R.

7 The polar curve C is given by the equation
$r = 6\cos 2\theta$ for $-\frac{1}{4}\pi \leqslant \theta \leqslant \frac{1}{4}\pi$

a Prove that C is symmetrical about the initial line.
You may assume any properties of the cosine function.

b Copy and complete the table. Give answers
to 1 decimal place where appropriate.
Hence sketch the curve C.

θ	0	$\frac{1}{16}\pi$	$\frac{2}{16}\pi$	$\frac{3}{16}\pi$	$\frac{4}{16}\pi$
$r = 6\cos 2\theta$					

c Find the exact polar coordinates of the point P where the
half-line $\theta = \frac{1}{6}\pi$ intersects C.

d Use the result that 'the angle in a semi-circle is a right angle'
to show that C is *not* a circle.

8 The polar curve C_1 is defined by the equation
$r^2 = 9\cos 2\theta$ for $-\frac{1}{4}\pi \leqslant \theta \leqslant \frac{1}{4}\pi$

a Copy and complete the table.
When not exact, give (positive) values
of r correct to 1 decimal place.

θ	0	$\frac{1}{16}\pi$	$\frac{2}{16}\pi$	$\frac{3}{16}\pi$	$\frac{4}{16}\pi$
r					

b Hence sketch, on one diagram, the curve C_1 and the half-lines $\theta = \pm\frac{1}{4}\pi$
You may assume the curve is symmetrical in the initial line.

The curve C_2 is defined by the polar equation $r^2 = 9\sin 2\theta$ for $0 \leqslant \theta \leqslant \frac{1}{2}\pi$.

c Show that if $P(r, \theta)$ is any point on C_1 then $Q\left(r, \theta + \frac{1}{4}\pi\right)$ is a point on C_2.

d Hence, on a separate diagram, sketch the curve C_2 and describe
the geometrical transformation which maps the graph of C_1 onto C_2.

e Hence write down the equation of the half-line about which C_2 is symmetrical.

FP2

You can convert between polar and cartesian coordinates by using trigonometric ratios and Pythagoras' theorem.

The diagrams show the same point P expressed in both polar and cartesian coordinates:

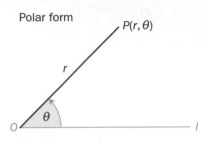

Polar form

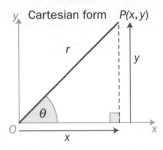

Cartesian form

To convert a polar equation into cartesian form, use the relationships

$$\cos\theta = \frac{x}{r}, \quad \sin\theta = \frac{y}{r}, \quad \tan\theta = \frac{y}{x}, \quad r^2 = x^2 + y^2$$

To convert a cartesian equation into polar form, use the relationships

$$x = r\cos\theta, \quad y = r\sin\theta, \quad x^2 + y^2 = r^2$$

Using trigonometry and Pythagoras' theorem.

EXAMPLE 1

A curve has polar equation $r = 8\cos\theta$

Show that this curve is a circle and state its radius.

A cartesian equation only involves terms in x and y.

Start with the polar equation $r = 8\cos\theta$

Replace $\cos\theta$ with $\frac{x}{r}$:

$$r = 8 \times \frac{x}{r} \quad \text{i.e.} \quad r^2 = 8x$$

Try to introduce a term in r^2.

Replace r^2 with $x^2 + y^2$ and rearrange the equation into a recognisable form:

The curve has cartesian equation $x^2 + y^2 - 8x = 0$

This is the equation of a circle.

Complete the square to find the radius of this circle:

$$(x - 4)^2 + y^2 = 16$$

The radius of this circle is 4.

See **C2** for revision. The circle $(x - a)^2 + (y - b)^2 = r^2$ has radius r, centre (a, b).

FP2

EXAMPLE 2

Find the polar equation of the curve given by $y = \frac{1}{4x}$

Replace y with $r\sin\theta$ and x with $r\cos\theta$ in the equation $y = \frac{1}{4x}$:

$$r\sin\theta = \frac{1}{4r\cos\theta}$$

i.e. $\quad r^2 = \dfrac{1}{4\sin\theta\cos\theta}$

$$= \frac{1}{2\sin 2\theta}$$

$$= \frac{1}{2}\operatorname{cosec} 2\theta$$

The curve $y = \frac{1}{4x}$ has polar equation $r^2 = \frac{1}{2}\operatorname{cosec} 2\theta$

See **C3** for revision.

$\sin 2\theta \equiv 2\sin\theta\cos\theta$

$\operatorname{cosec}\alpha \equiv \dfrac{1}{\sin\alpha}$

You do not need to make r the subject of your equation.

Exercise 7.3

1 Express these polar equations in cartesian form.
Sketch, on separate diagrams, the cartesian graph of each equation.

a $r = 4\sin\theta$

b $r = 5\cos\theta$

c $r = \dfrac{3}{2\cos\theta}$

d $r = 4\operatorname{cosec}\theta$

e $r = \dfrac{2}{\sin\theta + \cos\theta}$

f $r^2 = 2\operatorname{cosec} 2\theta$

2 Express these polar equations in cartesian form.
Give each answer in the form $y = f(x)$ and state any restriction on the values of x.

a $r = 4$

b $r^2 = 4$

c $\theta = \frac{1}{3}\pi$

d $\theta = -\frac{1}{6}\pi$

3 Express each polar equation in cartesian form, where y is a function of x. You do not need to make y the subject.

a $r = 2(\sin\theta - \cos\theta)$

b $r = \sqrt{2}\cos\left(\theta + \frac{1}{4}\pi\right)$

c $r^2\cos 2\theta = 1$

d $r^2 = \sin 2\theta$

e $r^2 = \tan\theta$

f $r^4 = 6\operatorname{cosec} 2\theta$

FP2

4 The diagram shows the polar curve C with equation
$r = 6\cos\theta$ for $0 \leqslant \theta \leqslant \frac{1}{2}\pi$.

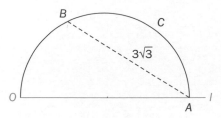

Point A has polar coordinates $A(6, 0)$.
Point B on the curve is such that $AB = 3\sqrt{3}$

a Verify that point A lies on C.

b By converting the equation $r = 6\cos\theta$ into cartesian form show that C is a semi-circle.

c Hence find the exact polar coordinates of point B.
Give your answer in the form (r, θ) for exact $0 \leqslant \theta \leqslant \frac{1}{2}\pi$.

5 Express these cartesian equations in polar form.
Give each answer in the form $r = f(\theta)$

a $x = 4$ **b** $2y = 1$ **c** $x^2 + y^2 = 2x$

d $x^2 + y^2 = 4y$ **e** $\dfrac{x^2 + y^2}{x + y} = 4$ **f** $(x + y)^2 = 1$

6 The diagram shows the polar curve C given by the equation
$r = 4a\sin\theta$ for $0 \leqslant \theta \leqslant \pi$, where a is a positive constant. O is the pole.

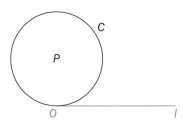

a By converting the equation $r = 4a\sin\theta$ to cartesian form, show that C is a circle.

b Write down, in terms of a, the coordinates of the centre P of this circle in
  i cartesian form   **ii** polar form.

c Find the polar coordinates of the points $Q(r, \theta)$ on C such that triangle OPQ is equilateral. Give r in terms of a and θ in terms of π, with $0 \leqslant \theta \leqslant \pi$.

7 The diagram shows the line L with cartesian equation $\sqrt{3}y + x = 6$

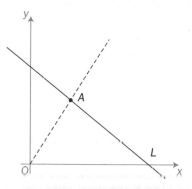

a Express $\sqrt{3}\sin\theta + \cos\theta$ in the form $k\cos(\alpha - \theta)$,
for $k > 0$ and α exact, $0 \leqslant \alpha \leqslant \frac{1}{2}\pi$.

b Hence show that the polar equation of L is given by

$$r = 3\sec\left(\frac{1}{3}\pi - \theta\right)$$

The line perpendicular to L passing through O intersects L at point A.

c By considering the minimum value attained by the expression $3\sec\left(\frac{1}{3}\pi - \theta\right)$, or otherwise, find

  i the distance OA

  ii the exact acute angle that the line OA makes with the positive x-axis

  iii the exact cartesian coordinates of A.

8 The diagram shows the polar curve C with equation $r^2\cos 2\theta = k$ for $0 \leqslant \theta < \frac{1}{4}\pi$ where k is a constant.

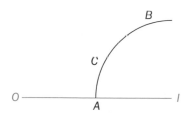

Point A on the curve has polar coordinates $(2, 0)$.

a Find the value of k.

b Show that a cartesian equation for C is $x^2 - y^2 = 4$

Point B on the curve is such that the distance $AB = 4$.

c Find the polar coordinates of point B. Give your answer in the form (r, θ) where r is in simplified surd form and θ is exact, $0 \leqslant \theta \leqslant \frac{1}{2}\pi$.

You can use differentiation to find the points of contact between a polar curve C and a tangent to C which is parallel or perpendicular to the initial line.

See **C1** for revision on tangents to curves.

In the diagram, the line perpendicular to l is a tangent to the polar curve $r = f(\theta)$ at point Q.

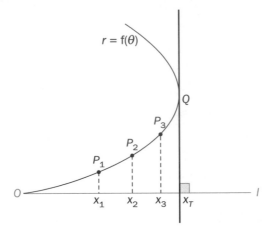

As θ varies and the point P moves along the curve, the x-coordinate of P approaches a maximum value x_T.

The tangent at Q can be located by solving the equation $\dfrac{dx}{d\theta} = 0$

To find the polar coordinates of the points on a curve at which a tangent is perpendicular to the initial line:

- express x in terms of θ and solve the equation $\dfrac{dx}{d\theta} = 0$.

To find the polar coordinates of the points on a curve at which a tangent is parallel to the initial line:

- express y in terms of θ and solve the equation $\dfrac{dy}{d\theta} = 0$.

x is a function of θ since $x = r\cos\theta$ and $r = f(\theta)$

EXAMPLE 1

The diagram shows the polar curve with equation $r = \sin\theta$ for $0 \leqslant \theta \leqslant \frac{1}{2}\pi$.

The tangents to the curve at points P and Q are, respectively, parallel and perpendicular to the initial line, as shown. Find the polar coordinates of

a point P

b point Q.

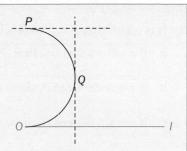

---

a Write y as a function of θ using the curve equation $r = \sin\theta$:

$$y = r\sin\theta \quad \text{so} \quad y = \sin^2\theta$$

The aim is to solve the equation $\dfrac{dy}{d\theta} = 0$

See Section 7.3. $y = r\sin\theta$ for any polar curve.

Differentiate y using the chain rule to find $\dfrac{dy}{d\theta}$:

$$\frac{dy}{d\theta} = 2\sin\theta\cos\theta$$
$$= \sin 2\theta$$

Use trigonometric identities to simplify equations. Refer to **C3**.

Solve $\dfrac{dy}{d\theta} = 0$ to find the points where the tangent is parallel to the initial line:

For $0 \leqslant \theta \leqslant \frac{1}{2}\pi$, the equation $\sin 2\theta = 0$ has solution $\theta = 0$ or $\theta = \frac{1}{2}\pi$.

For $0 \leqslant 2\theta \leqslant \pi$
$\sin 2\theta = 0$ for $2\theta = 0,\ \pi$

Since $\theta = 0$ gives the initial line, point P corresponds to $\theta = \frac{1}{2}\pi$.

Use the polar equation of the curve $r = \sin\theta$ to find the value of r at P:

$$r = \sin\left(\frac{1}{2}\pi\right) = 1$$

i.e. P has polar coordinates $\left(1, \frac{1}{2}\pi\right)$

b Write x as a function of θ:

$$x = r\cos\theta \quad \text{so} \quad x = \sin\theta\cos\theta$$
$$= \frac{1}{2}\sin 2\theta$$

The aim is to solve the equation $\dfrac{dx}{d\theta} = 0$

Differentiate x using the chain rule to find $\dfrac{dx}{d\theta}$:

$$\frac{dx}{d\theta} = \cos 2\theta$$

Solve $\dfrac{dx}{d\theta} = 0$ to find the points where the tangent is perpendicular to the initial line:

For $0 \leqslant \theta \leqslant \frac{1}{2}\pi$, the equation $\cos 2\theta = 0$ has solution $\theta = \frac{1}{4}\pi$.

For $0 \leqslant 2\theta \leqslant \pi$
$\cos 2\theta = 0$ for $2\theta = \frac{1}{2}\pi$

Hence Q has polar coordinates $\left(\frac{1}{2}\sqrt{2}, \frac{1}{4}\pi\right)$.

$r = \sin\left(\frac{1}{4}\pi\right) = \frac{1}{2}\sqrt{2}$

FP2

Exercise 7.4

1 It is given that $x = r\cos\theta$, where r is a function of θ.

    a If $r = 3\cos\theta$, show that $\dfrac{dx}{d\theta} = -3\sin 2\theta$

    b If $r = \sec\theta + \tan\theta$, show that $\dfrac{dx}{d\theta} = \cos\theta$

    c If $r = \sin\theta + \cos\theta$, show that $\dfrac{dx}{d\theta} = \cos 2\theta - \sin 2\theta$

2 It is given that $y = r\sin\theta$, where r is a function of θ.

    a If $r = \cos^2\theta$, show that $\dfrac{dy}{d\theta} = \cos\theta(1 - 3\sin^2\theta)$

    b If $r = \sec\theta - \operatorname{cosec}\theta$, show that $\dfrac{dy}{d\theta} = \sec^2\theta$

    c If $r = \dfrac{1}{\sin\theta + \cos\theta}$, show that $\dfrac{dy}{d\theta} = \dfrac{1}{(\sin\theta + \cos\theta)^2}$

3 The diagram shows the curve C with polar equation
$r = 4\cos\theta$ where $-\dfrac{1}{2}\pi \leqslant \theta \leqslant \dfrac{1}{2}\pi$.

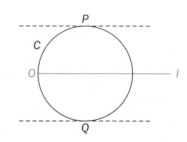

The tangents to the curve at points P and Q are parallel to the initial line.

    a Use calculus to find the polar coordinates of the points P and Q.
Give each answer in exact form.

    b Find the area of triangle OPQ.

4 The diagram shows the polar curve with equation
$r = \sin^2\theta \quad$ for $0 \leqslant \theta \leqslant \dfrac{1}{2}\pi$.

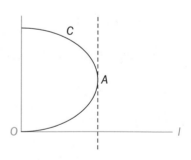

The tangent to the curve at point A is perpendicular to the initial line.

    a Show that the x-coordinate of any point $P(r, \theta)$ on the curve is given by $x = \cos\theta - \cos^3\theta$

    b Find $\dfrac{dx}{d\theta}$.

    c Hence show that $OA = \dfrac{2}{3}$

5 The diagram shows the curve with polar equation
$r = 1 - \cos\theta$ for $0 \leqslant \theta \leqslant \pi$.
Tangents to the curve at points A and B are, respectively, perpendicular and parallel to the initial line l.

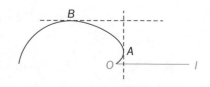

    a Show that A has polar coordinates $\left(\dfrac{1}{2}, \dfrac{1}{3}\pi\right)$.

    b Find the exact polar coordinates of point B and hence show that the line AB has length $\dfrac{1}{2}\sqrt{7}$.

6 The diagram shows the polar curve defined by the equation
$r = 4(\cos\theta - \sin\theta)$ for $-\frac{1}{4}\pi \leqslant \theta \leqslant \frac{1}{4}\pi$.

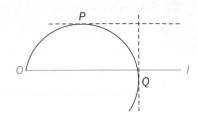

The tangents to the curve at points P and Q are, respectively, parallel and perpendicular to the initial line l.

a Use the relationship $y = r\sin\theta$ to show that, for any point (r, θ) on the curve, $y = 2(\sin 2\theta + \cos 2\theta - 1)$

Point P has polar coordinates $P(r_1, \theta_1)$ where $-\pi < \theta_1 \leqslant \pi$.

b Show that $\theta_1 = \frac{1}{8}\pi$.

It is given that point Q has polar coordinates $Q\left(r_2, -\frac{1}{8}\pi\right)$.

c Show that $r_1 r_2 = 8\sqrt{2}$ and hence find the area of triangle OPQ.

7 The diagram shows the polar curve with equation $r = \dfrac{1}{2 - \sin\theta}$ for $-\pi < \theta \leqslant \pi$.

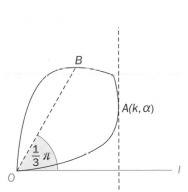

The curve is enclosed by the rectangle $ABCD$ whose sides form tangents to the curve, which are either parallel or perpendicular to the initial line l, as shown.

a Find the exact polar coordinates of the points where the tangents AB and DC each intersect this curve.

b Find the exact polar coordinates of the points where the tangents AD and BC each intersect this curve.

c Show that the area of rectangle $ABCD$ is $\frac{8}{9}\sqrt{3}$.

8 The diagram shows the polar curve defined by the equation
$r^2 = 3\sin 2\theta$ for $0 \leqslant \theta \leqslant \frac{1}{2}\pi$.

The tangent to the curve perpendicular to l touches the curve at point A, where A has polar coordinates $A(k, \alpha)$, $k > 0$, $0 < \alpha < \frac{1}{2}\pi$.

a Use the relationship $x = r\cos\theta$ to show that for any polar point (r, θ) on the curve $x^2 = 6\sin\theta\cos^3\theta$

b Using implicit differentiation, or otherwise, show that $\alpha = \frac{1}{6}\pi$.

Point B on the curve is such that the acute angle between OB and the line l is $\frac{1}{3}\pi$.

c Find the area of triangle OAB. Give your answer in the form $p\sqrt{q}$ where p is a rational number and q is an integer to be stated.

FP2

159

Intersection of two polar curves

You can find the points where the polar curves $r = f(\theta)$ and $r = g(\theta)$ intersect by solving the equation $f(\theta) = g(\theta)$.

The diagram shows the polar curves with equations $r = f(\theta)$ and $r = g(\theta)$. The curves intersect at a point $P(r_0, \theta_0)$.

Since P lies on the curve,
$r = f(\theta)$, $r_0 = f(\theta_0)$

Similarly $r_0 = g(\theta_0)$

Hence $f(\theta_0) = g(\theta_0)$

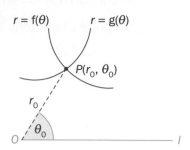

The coordinates of the points where the polar curves with equations $r = f(\theta)$ and $r = g(\theta)$ intersect are given by (r_0, θ_0) where $r_0 = f(\theta_0)$ and θ_0 satisfies the equation $f(\theta) = g(\theta)$

EXAMPLE 1

The diagram shows the polar curves C_1 and C_2 with polar equations

$$C_1: r = \sqrt{2}\sin\theta \quad 0 \leqslant \theta \leqslant \tfrac{1}{2}\pi$$

and $\quad C_2: r = \sqrt{2}\cos\theta \quad 0 \leqslant \theta \leqslant \tfrac{1}{2}\pi$

Find the polar coordinates of the point P where the two curves intersect.

Solve $f(\theta) = g(\theta)$ to find any points of intersection:

$f(\theta) = \sqrt{2}\sin\theta, \quad g(\theta) = \sqrt{2}\cos\theta$

$$\sqrt{2}\sin\theta = \sqrt{2}\cos\theta$$

$$\text{so} \quad \tan\theta = 1 \quad \text{i.e.} \quad \theta = \tfrac{1}{4}\pi$$

The equation $\sqrt{2}\sin\theta = \sqrt{2}\cos\theta$

implies $\cos\theta \neq 0$ and so $\dfrac{\sin\theta}{\cos\theta} \equiv \tan\theta$

For $0 \leqslant \theta \leqslant \tfrac{1}{2}\pi$, $\tan\theta = 1$ only

when $\theta = \tfrac{1}{4}\pi$

Use the curve equation $C_1: r = \sqrt{2}\sin\theta$ to find the r-value at point P:

$$r = \sqrt{2}\sin\left(\tfrac{1}{4}\pi\right) = 1$$

The two curves intersect at the point $P\left(1, \tfrac{1}{4}\pi\right)$.

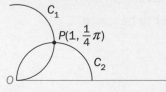

You could also use the equation $C_2: r = \sqrt{2}\cos\theta$ to find r.

Although both curves pass through O, they do so for *different* values of θ (for C_1 when $\theta = 0$, and for C_2 when $\theta = \tfrac{1}{2}\pi$) so the pole O is not a point of intersection.

Exercise 7.5

1 Find the exact polar coordinates of the point of intersection of these polar curves.

a $r = \sin\theta, r = \sqrt{3}\cos\theta$ for $0 \leqslant \theta < \frac{1}{2}\pi$

b $r = 2\cos\theta$ and $r = \sqrt{12}\sin\theta$ for $0 \leqslant \theta < \frac{1}{2}\pi$

c $r = 2 + 3\sin\theta$ and $r = 6 - \sin\theta$ for $0 \leqslant \theta \leqslant \pi$

d $r = 3 + 2\cos\theta$ and $r = 5 - 2\cos\theta$ for $0 \leqslant \theta \leqslant 2\pi$

2 Find the exact polar coordinates of the point(s) of intersection of these polar curves.

a $r = 2\sin\theta$ and $r = \operatorname{cosec}\theta$ for $0 < \theta < \pi$

b $r = 2\cos^2\theta$ and $r = 3\cos\theta - 1$ for $0 \leqslant \theta \leqslant 2\pi$

c $r = \cos 2\theta$ and $r = \cos\theta$ for $-\frac{1}{4}\pi \leqslant \theta \leqslant \frac{1}{4}\pi$

d $r = \cos 2\theta$ and $r = 3\sin\theta - 1$ for $\frac{1}{6}\pi \leqslant \theta \leqslant \frac{1}{4}\pi$

3 The diagram shows the polar curves C_1 and C_2 with equations given by

$$C_1: r = \sin 2\theta \quad 0 \leqslant \theta \leqslant \frac{1}{2}\pi$$

and $C_2: r = \sin\theta \quad 0 \leqslant \theta \leqslant \frac{1}{2}\pi$

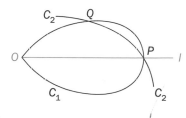

a Verify that the two curves intersect at the pole O.

b Find the exact coordinates of the other point where these curves meet.

4 The diagram shows the polar curves C_1 and C_2 with equations given by

$$C_1: r = \cos 2\theta \quad -\frac{1}{4}\pi \leqslant \theta \leqslant \frac{1}{4}\pi$$

and $C_2: r = 1 - \sin\theta \quad -\frac{1}{4}\pi \leqslant \theta \leqslant \frac{1}{4}\pi$

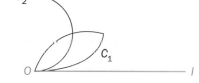

a By using a suitable trigonometric identity, find the exact coordinates of the points P and Q where these curves intersect.

b Calculate the area of triangle OPQ.

5 The diagram shows the circle C with equation $r = 4\cos\theta$ for $-\frac{1}{2}\pi \leqslant \theta \leqslant \frac{1}{2}\pi$ and the straight line L with equation

$r = 3\sec\theta$ for $-\frac{1}{2}\pi < \theta < \frac{1}{2}\pi$.

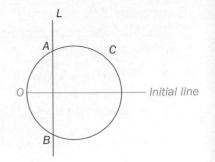

The line L intersects the circle C at the points A and B.

a Find the exact polar coordinates of the point A and the point B.

b Hence show that the chord AB of the circle has length $2\sqrt{3}$.

6 The polar curves C_1 and C_2 have equations given by

$$C_1: r = 2\sin^2\theta \qquad 0 \leqslant \theta \leqslant \frac{1}{2}\pi$$

and $C_2: r = 3(1 - \cos\theta) \quad 0 \leqslant \theta \leqslant \frac{1}{2}\pi$

a Find the polar coordinates of the points of intersection of these curves.

b By considering the polar coordinates of the points on the curve when $\theta = \frac{1}{2}\pi$, or otherwise, state, with a reason, which of these diagrams best represents the graphs of C_1 and C_2.

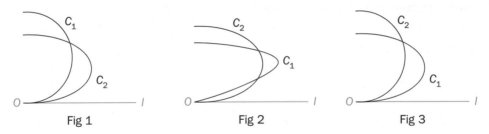

Fig 1   Fig 2   Fig 3

7 The diagram shows the polar curves C_1 and C_2 with equations given by

$$C_1: r = 2\sin^2\theta \qquad 0 \leqslant \theta \leqslant \pi$$

and $C_2: r = 1 + k\cos\theta \quad 0 \leqslant \theta \leqslant \pi$

where k is a constant.

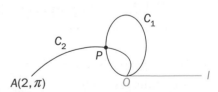

Point P shown on the diagram is a point of intersection of the two curves.

a Given that the point $A(2, \pi)$ lies on C_2, show that $k = -1$.

b Find the exact polar coordinates of the points where these curves intersect.

c Show that $AP = \frac{1}{2}\sqrt{13}$

d Find the polar coordinates of the point Q which is the midpoint of the line AP. Give your answer in the form (r, θ) where r is exact and θ is in radians, correct to 1 decimal place.

8

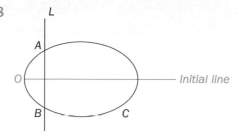

The diagram shows the polar curve C and straight line L with equations given by

$$C: r = 1 + \cos 2\theta \quad -\tfrac{1}{2}\pi \leqslant \theta \leqslant \tfrac{1}{2}\pi$$

and $\quad L: r = \dfrac{1}{4}\sec\theta \quad -\tfrac{1}{2}\pi < \theta < \tfrac{1}{2}\pi$

The curve C and the line L intersect at points A and B as shown.

a Find the exact polar coordinates of point A and of point B.

b Show that the area of the triangle ABD, where D is the point on this curve which is furthest from O, is $\dfrac{7}{16}\sqrt{3}$.

9 The diagrams shows the circle C with equation $r = 4\sin\theta$ for $0 \leqslant \theta \leqslant \pi$ and the straight line L with equation $r = \sqrt{3}\sec\theta$ for $0 \leqslant \theta < \tfrac{1}{2}\pi$. Line L intersects circle C at points A and B.

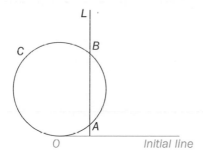

a Find the cartesian equation of the circle C and hence state the radius of C.

b Find the exact polar co-ordinates of the point A and the point B.

c Show that line AB has length 2 units.

d Find the exact area of the region bounded by the lines OA and OB and the minor arc AB of circle C.

Area of a sector of a polar curve

You can use integration to find the area of a region swept out by a polar curve.

The diagram shows part of a polar curve with equation $r = f(\theta)$, where θ ranges from $\theta = \alpha$ to $\theta = \beta$ for fixed values $\alpha < \beta$.

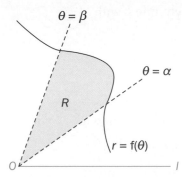

The area A of region R bounded by the curve $r = f(\theta)$ and the half-lines $\theta = \alpha$ and $\theta = \beta$ is

$$A = \frac{1}{2} \int_{\alpha}^{\beta} r^2 \, d\theta$$

given that $r \geqslant 0$ for all values of θ where $\alpha \leqslant \theta \leqslant \beta$.

This result is in the **FP2** section of the formula book.

EXAMPLE 1

The diagram shows a semi-circle with polar equation $r = 2\cos\theta$ for $0 \leqslant \theta \leqslant \frac{1}{2}\pi$.

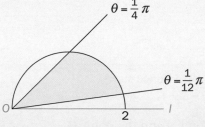

a Find the area of the region bounded by the curve and the half-lines $\theta = \frac{1}{12}\pi$ and $\theta = \frac{1}{4}\pi$, shaded in the diagram. Give your answer to 2 decimal places.

b Use calculus to show that the area of the semi-circle is $\frac{1}{2}\pi$.

In the exam, you must use the method specified.

EXAMPLE 1 (CONT.)

a Use the formula to find A, the area swept out by the curve between the two given lines:

$$A = \frac{1}{2}\int_{\frac{\pi}{12}}^{\frac{\pi}{4}} r^2\, d\theta = \frac{1}{2}\int_{\frac{\pi}{12}}^{\frac{\pi}{4}} (2\cos\theta)^2\, d\theta$$

Replace r with $2\cos\theta$.

$$= \int_{\frac{\pi}{12}}^{\frac{\pi}{4}} 2\cos^2\theta\, d\theta$$

$\frac{1}{2}\times(2\cos\theta)^2 = \frac{1}{2}\times 4\cos^2\theta = 2\cos^2\theta$

$$= \int_{\frac{\pi}{12}}^{\frac{\pi}{4}} \cos 2\theta + 1\, d\theta$$

See **C3** for revision.
Double-angle identity: $\cos 2\theta \equiv 2\cos^2\theta - 1$

$$= \left[\frac{1}{2}\sin 2\theta + \theta\right]_{\frac{\pi}{12}}^{\frac{\pi}{4}}$$

$\int \cos k\theta\, d\theta = \frac{1}{k}\sin k\theta + c$ for constant k.

$$= \left(\frac{1}{2} + \frac{1}{4}\pi\right) - \left(\frac{1}{4} + \frac{1}{12}\pi\right) = \frac{1}{4} + \frac{1}{6}\pi$$

$\sin\left(\frac{1}{2}\pi\right) = 1$, $\sin\left(\frac{1}{6}\pi\right) = \frac{1}{2}$

The required area $A = 0.77$ to 2 decimal places.

The *exact* area $A = \frac{1}{4} + \frac{1}{6}\pi = 0.7735\ldots$

b The semi-circle is the region bounded by the curve and the half-lines $\theta = 0$ and $\theta = \frac{1}{2}\pi$.

The initial line l has equation $\theta = 0$.

The area of the semi-circle is given by $\frac{1}{2}\int_0^{\frac{\pi}{2}} r^2\, d\theta$

Use the integral from **a**:

$$\frac{1}{2}\int_0^{\frac{\pi}{2}} (2\cos\theta)^2\, d\theta = \left[\frac{1}{2}\sin 2\theta + \theta\right]_0^{\frac{\pi}{2}}$$

$$= \left(\frac{1}{2}\times 0 + \frac{1}{2}\pi\right) - \left(\frac{1}{2}\times 0 + 0\right)$$

$$= \frac{1}{2}\pi \text{ as required.}$$

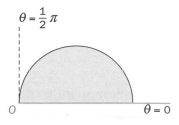

The answer makes sense since the area of a semi-circle of unit radius $= \frac{1}{2}\pi \times 1^2 = \frac{1}{2}\pi$

FP2

Exercise 7.6

1 Find the exact area bounded by these polar curves with equation $r = f(\theta)$ and the given half-lines.

   a $r = 2\sin\theta,\ \theta = \frac{1}{4}\pi,\ \theta = \frac{1}{2}\pi$ **b** $r = \sqrt{4\cos 2\theta},\ \theta = \frac{1}{6}\pi,\ \theta = \frac{1}{4}\pi$

   c $r = \sin\theta,\ \theta = \frac{1}{2}\pi,\ \theta = \pi$ **d** $r = 2 + 2\sin\theta,\ \theta = 0,\ \theta = \frac{1}{2}\pi$

2 A polar curve is defined by the equation $r = 2\cos 2\theta$ for $0 \leqslant \theta \leqslant \frac{1}{4}\pi$.

   a Show that $r^2 = 2(\cos 4\theta + 1)$

   b Hence find the exact area bounded by this curve and the half-lines $\theta = \frac{1}{24}\pi$ and $\theta = \frac{1}{8}\pi$.

3 The diagram shows the polar curve defined by the equation
$r = 2(2 - \sin\theta)$ for $0 \leqslant \theta \leqslant \frac{1}{2}\pi$, and the half-lines $\theta = 0$
and $\theta = \frac{1}{2}\pi$.

Show that area of the region enclosed by the curve and the
given half-lines is $\frac{1}{2}(9\pi - 16)$.

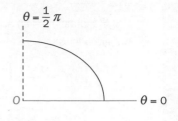

4 Find the exact area bounded by these polar curves with equation
$r = f(\theta)$ and the given half-lines.

   a $r = \sqrt{2\theta + 1},\ \theta = 0,\ \theta = 4$ b $r = \frac{\pi}{\theta},\ \theta = \pi,\ \theta = 2\pi$

   c $r = 2e^{2\theta},\ \theta = 0,\ \theta = 2\pi$ d $r = \dfrac{1}{\sqrt{2\theta + e}}\ \theta = 0,\ \theta = e$

5 The diagram shows the polar curve with equation
$r = 2a(\cos\theta + \sin\theta)$, where a is a positive constant, for $0 \leqslant \theta \leqslant \frac{3}{4}\pi$.

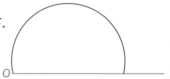

   a Use trigonometric identities to show that, for this
      polar curve, $r^2 = 4a^2(1 + \sin 2\theta)$

   b Find the exact total area enclosed by the curve.
      Give your answer in terms of a.

6 The diagram shows the straight line L given by the polar
equation $r = 2\sec\theta$.

The half-lines $\theta = \frac{1}{4}\pi$ and $\theta = \frac{1}{3}\pi$ intersect L at points
P and Q respectively.

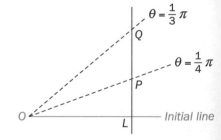

   a Use calculus to show that the area of triangle OPQ
      is $2(\sqrt{3} - 1)$.

   b Find the polar coordinates of points P and Q.

   c Hence, by finding an alternative expression for the area of
      triangle OPQ, show that $\sin\left(\frac{1}{12}\pi\right) = \dfrac{1}{2\sqrt{2}}(\sqrt{3} - 1)$

7 The diagram shows the circle C with polar equation $r = 4\sin\theta$
for $0 \leqslant \theta \leqslant \pi$. The half-lines $\theta = \frac{1}{6}\pi$ and $\theta = \frac{1}{3}\pi$ intersect
the circle at points P and Q respectively.

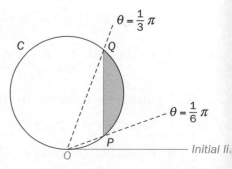

   a Find the polar coordinates of point P and point Q.

   b Find the exact area of triangle OPQ.

   c Use calculus to show that the area of the region bounded
      by the line PQ and the circle, as shaded on the diagram, is
      given by $\frac{1}{3}(2\pi - 3\sqrt{3})$.

FP2

8 The diagram shows the polar curves C_1 and C_2 with equations given by

$$C_1: r_1 = 2 - \sqrt{2}\cos\theta \quad 0 \leqslant \theta \leqslant \tfrac{1}{2}\pi$$

$$C_2: r_2 = \sqrt{2}\cos\theta \quad 0 \leqslant \theta \leqslant \tfrac{1}{2}\pi$$

and the half-lines $\theta = 0$ and $\theta = \tfrac{1}{2}\pi$.

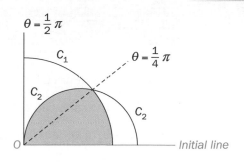

The half-line $\theta = \tfrac{1}{4}\pi$ passes through the point where these curves intersect.

Show that the area of the finite region between these curves shaded on the diagram is $\tfrac{1}{4}(3\pi - 8)$.

9 The diagram shows the polar curves C_1 and C_2 with equations given by

$$C_1: r_1 = 2 - \sin\theta \quad 0 \leqslant \theta \leqslant \tfrac{1}{2}\pi$$

$$C_2: r_2 = 3\cos 2\theta \quad 0 \leqslant \theta \leqslant \tfrac{1}{4}\pi$$

and the initial lines $\theta = 0$ and $\theta = \tfrac{1}{2}\pi$.

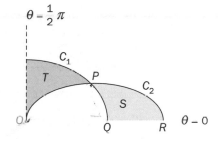

The curves intersect at point P.
Points Q on C_1 and R on C_2 lie on the initial line.

a Show that the polar coordinates of point P are $\left(\tfrac{3}{2}, \tfrac{1}{6}\pi\right)$

b Hence find the area of region S bounded by these curves and the initial line, and which contains the points P, Q and R. Give your answer to 2 decimal places.

T is the region bounded by these curves and the half-line $\theta = \tfrac{1}{2}\pi$.

c **i** Explain why the integral $\tfrac{1}{2}\displaystyle\int_{\frac{\pi}{6}}^{\frac{\pi}{2}} \left(r_1^2 - r_2^2\right)\,\mathrm{d}\theta$ does *not* calculate the area of T.

ii Use calculus to show that the exact area of T is given by $\tfrac{1}{32}(18\pi - 21\sqrt{3})$.

1 A curve C is defined by the polar equation $r^2\sin 2\theta = 10$ for $0 < \theta < \frac{1}{2}\pi$.

   a   Find the cartesian equation of this curve.

   b   Hence, or otherwise, sketch the cartesian graph of C.

2 **a**   Express $\sin\theta - \cos\theta$ in the form $k\sin(\theta - \alpha)$ for exact $k > 0, 0 \leqslant \alpha \leqslant \frac{1}{2}\pi$.

   b   Hence, or otherwise, find the polar equation of the line given by the cartesian equation $y - x = 4\sqrt{2}$

3 The polar curves C_1 and C_2 are defined by the equations

$$C_1: r = 2a\sin^2\theta \qquad \text{for } 0 \leqslant \theta \leqslant \frac{1}{2}\pi$$

$$C_1: r = a(2 - \cos\theta) \quad \text{for } 0 \leqslant \theta \leqslant \frac{1}{2}\pi$$

where a is a positive constant.

   a   Find the exact polar coordinates of the points where the two curves intersect. Give your answers in terms of a.

   b   Show that the area of triangle OPQ is $\frac{3}{4}a^2$, where P and Q are these points of intersection.

4 The diagram shows the polar curve with equation
$r = 3\sin^2\theta$ for $0 \leqslant \theta \leqslant \frac{1}{2}\pi$.

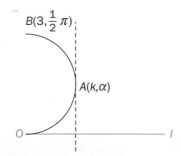

A tangent to the curve, perpendicular to the initial line l, touches the curve at point A with polar coordinates (k, α) where $k > 0$ and $0 < \alpha < \frac{1}{2}\pi$.

Point B on the curve has polar coordinates $\left(3, \frac{1}{2}\pi\right)$.

   a   Show that $\dfrac{\mathrm{d}x}{\mathrm{d}\theta} = 3\sin\theta\left(3\cos^2\theta - 1\right)$

   b   Hence find the exact value of $\cos\alpha$ and the exact value of k.

   c   Show that triangle OAB has area $\sqrt{3}$.

FP2

5 The diagram shows a curve C with polar equation $r = 2 + \sqrt{2}a\sin\theta$ for $0 \leqslant \theta \leqslant 2\pi$, where $0 < a < \sqrt{2}$, and the half-lines $\theta = 0$, $\theta = \frac{1}{2}\pi$, which intersect the curve at points H and K respectively.

The region R, shown shaded on the diagram, is bounded by C and the line HK and does not contain the pole O.

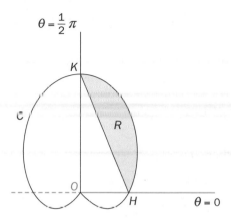

a Show that area of the region R is $\frac{1}{4}\pi(4 + a^2) + \sqrt{2}a - 2$.

b Find, in terms of a and π, the total area enclosed by this curve.

c Deduce that R occupies less than one-quarter of this total area.

6 The polar curve C and line L are defined by the equations

$$C: r = 4\sqrt{3}\cos 2\theta$$
$$L: r = 3\sec\theta$$

where $0 \leqslant \theta \leqslant \frac{1}{4}\pi$.

a Verify that the two curves intersect at the point with polar coordinates $P\left(2\sqrt{3}, \frac{1}{6}\pi\right)$.

b On the same diagram sketch the graphs of C and L. Show clearly the position of point P.

c Find the exact area enclosed by the curve C.

d Show that the line L divides this area in the ratio $1:2$.

7 The diagram shows the curve C defined by the polar equation

$C: r = 2\cos\theta + 2 \quad$ for $0 \leqslant \theta \leqslant \pi$

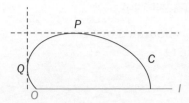

Tangents to the curve parallel and perpendicular to the initial line l touch the curve at points P and Q respectively, as shown.

a Show that P has polar coordinates $\left(3, \frac{1}{3}\pi\right)$.

b Find the exact polar coordinates of point Q.
Give your answer in the form (k, α) where $0 \leqslant \alpha \leqslant \pi$ is exact.

c Show that $PQ = \sqrt{7}$.

d Find the exact area of the finite region bounded by the curve and the lines OP and OQ.

8 The diagram shows the curve C given by the polar equation

$C: r = 2a\sin 2\theta$ for $0 \leqslant \theta \leqslant \frac{1}{2}\pi$, where $a > 0$. The half-lines $\theta = \frac{1}{6}\pi$ and $\theta = \frac{1}{3}\pi$ intersect C at the points A and B respectively.

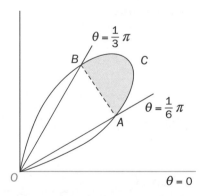

a Find, in terms of a, the area of triangle OAB.

b Use calculus to show that the area of the region bounded by the curve and the line AB, as shaded on the diagram, is given by

$$\frac{1}{12}a^2\left(2\pi + 3\sqrt{3} - 9\right)$$

9 The diagram shows a curve C defined by the polar equation

$C: r = a(1 - \sin\theta)$ for $0 \leqslant \theta \leqslant \frac{1}{2}\pi$, where $a > 0$. The line L is parallel to the initial line and is a tangent to C at point A.

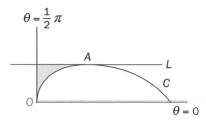

a Find the polar coordinates of point A. Give your answer in exact form and in terms of a.

b Hence show that the area of the region bounded by C, the line L and the half-line $\theta = \frac{1}{2}\pi$, shown shaded on the diagram, is given by

$$\frac{1}{32}a^2\left(15\sqrt{3} - 8\pi\right)$$

10 The diagram shows a curve C defined by the polar equation

$C: r^2 = a^2\cos 2\theta$ for $-\frac{1}{4}\pi \leqslant \theta \leqslant \frac{1}{4}\pi$, where $a > 0$. Tangents to the curve, parallel to the initial line, intersect C at points A and B as shown.

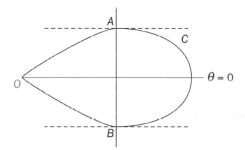

a Show that if $P(r, \theta)$ is any polar point on C then

$$r^2\sin^2\theta = a^2(\sin^2\theta - 2\sin^4\theta)$$

b Use differentiation to find the polar coordinates of point A and point B. Give answers in terms of a and in exact form.

c Show that the line AB divides the total area enclosed by the curve in the ratio

$$4 - \sqrt{3} : \sqrt{3}$$

d Write down, in terms of a, the polar equation of the line AB.

7 Exit ⟹

Summary

Refer to

- The **polar coordinates** of a general point P are (r, θ) where $r \geqslant 0$ is the length of the line OP and θ is an angle, in radians, between the line OP and the **initial line**. θ is negative if measured in a clockwise direction.

  7.1

- To convert a polar equation into **cartesian form**, use the relationships
  $$\cos \theta = \frac{x}{r}, \quad \sin \theta = \frac{y}{r}, \quad \tan \theta = \frac{y}{x}, \quad r^2 = x^2 + y^2$$

  7.3

- To convert a cartesian equation into polar form, use the relationships
  $$x = r\cos \theta, \quad y = r\sin \theta, \quad x^2 + y^2 = r^2$$

  7.3

- To find the polar coordinates of the points at which the tangent of a curve is

  - perpendicular to the initial line: solve the equation $\dfrac{dx}{d\theta} = 0$, where $x = f(\theta)\cos \theta$
  - parallel to the initial line: solve the equation $\dfrac{dy}{d\theta} = 0$, where $y = f(\theta)\sin \theta$.

  7.4

- The coordinates of the points where the polar curves with equations $r = f(\theta)$ and $r = g(\theta)$ intersect are given by (r_0, θ_0) where $r = f(\theta_0)$ and θ_0 satisfies the equation $f(\theta) = g(\theta)$

  7.5

- The area A of region R bounded by the curve $r = f(\theta)$ and the half-lines $\theta = \alpha$ and $\theta = \beta$ is given by $A = \dfrac{1}{2}\displaystyle\int_{\alpha}^{\beta} r^2 \, d\theta$

  provided $r \geqslant 0$ for all $\alpha \leqslant \theta \leqslant \beta$.

  7.6

Links

Polar coordinates were initially invented to describe circular and orbital motion and are often used in navigation.

Systems involving radial symmetry, radial forces, or radial asymmetry can be modelled using polar coordinates in 2-dimensions, for example to describe the pickup pattern of a microphone.

In 3-dimensions, a number of polar plots can be combined to model the output of a loudspeaker.

FP2

1 a Find the Taylor expansion of $\cos 2x$ in ascending powers of $\left(x - \frac{1}{4}\pi\right)$ up to and including the term in $\left(x - \frac{1}{4}\pi\right)^5$.

b Use your answer to **a** to obtain an estimate of $\cos 2$, giving your answer to 6 decimal places. [(c) Edexcel Limited 2006]

2 The equation of a polar curve C is given by $r^2(1 - \sin 2\theta) = 4$, $0 \leqslant \theta < \frac{1}{4}\pi$

a Write down the polar coordinates of the point where C intersects the initial line.

b Find the cartesian equation of C and hence sketch its cartesian graph.

3 $\dfrac{dy}{dx} - y \tan x = 2\sec^3 x$

Given that $y = 3$ at $x = 0$, find y in terms of x. [(c) Edexcel Limited 2007]

4 a Find the general solution of the differential equation

$$2\frac{d^2y}{dt^2} + 7\frac{dy}{dt} + 3y = 3t^2 + 11t$$

b Find the particular solution of this differential equation for which $y = 1$ and $\dfrac{dy}{dt} = 1$ when $t = 0$.

c For this particular solution, calculate the value of y when $t = 1$. [(c) Edexcel Limited 2002]

5 a Write down the first three non-zero terms, in ascending powers of x, of the Maclaurin expansion of $\ln(1 + x)$.

b Deduce that, for sufficiently small x, $\ln(1 - x) \approx -x - \dfrac{x^2}{2} - \dfrac{x^3}{3}$

and state the range of values of x for which the full expansion is valid.

c Hence, or otherwise, find in ascending powers of x the Maclaurin expansion of $\ln\left(\dfrac{1 + x}{1 - x}\right)$ up to and including the term in x^3.

d By using a suitable value of x in the answer to **c** show that $\ln 3 \approx \dfrac{13}{12}$

FP2

6 The diagram shows the polar curves

$C_1: r = \sqrt{3}\sin 2\theta, \quad 0 \leqslant \theta \leqslant \frac{1}{2}\pi$

$C_2: r = 3\cos 2\theta, \quad 0 \leqslant \theta \leqslant \frac{1}{4}\pi.$

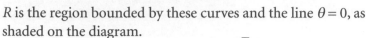

Point P is where the two curves intersect.

a Find the polar coordinates of point P.
Give your answer in exact form.

R is the region bounded by these curves and the line $\theta = 0$, as shaded on the diagram.

b Show that the area of region R is $\dfrac{3\sqrt{3}}{8} + \dfrac{1}{4}\pi$.

7 $(2xy + 1)\dfrac{dy}{dx} = x + y^2$ where $y = \dfrac{1}{2}$ at $x = 0$.

a Show that $(2xy + 1)\dfrac{d^2y}{dx^2} = 1 - 2x\left(\dfrac{dy}{dx}\right)^2$

b Hence find a series solution of this differential equation in ascending powers of x up to and including the term in x^2.

c Use this series approximation to estimate the value of y when $x = \dfrac{1}{2}$.

8 $\dfrac{d^2y}{dx^2} + 4\dfrac{dy}{dx} + 5y = 65\sin 2x, \quad x > 0$

a Find the general solution of the differential equation.

b Show that for large values of x this general solution may be approximated by a sine function and find this sine function.

[(c) Edexcel Limited 2004]

9 The diagram shows the curve C_1 which has polar equation

$r = a(3 + 2\cos\theta), \quad 0 \leqslant \theta < 2\pi$, and the circle C_2 with equation $r = 4a$, where a is a positive constant.

a Find, in terms of a, the polar coordinates of the points where the curve C_1 meets the circle C_2.

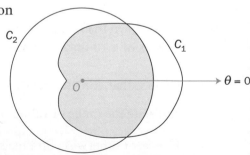

The regions enclosed by the curves C_1 and C_2 overlap and this common region R is shaded in the figure.

b Find, in terms of a, an exact expression for the area of the shaded region R.

c In a single diagram, copy the curves, C_1 and C_2, and also sketch the curve C_3 with polar equation $r = 2a\cos\theta, 0 \leqslant \theta < 2\pi$. Show clearly the coordinates of the points of intersection of C_1, C_2 and C_3 with the initial line $\theta = 0$.

[(c) Edexcel Limited 2008]

10 $f(x) = \ln(3 - 2x), \quad x < \frac{2}{3}$

   a Prove by induction that

$$f^{(n)}(x) = -(n - 1)!\,2^n(3 - 2x)^{-n}$$

   for all positive integers n.

   b Hence find the coefficient of $\left(x - \frac{1}{2}\right)^n$, $n \geqslant 1$, in the

   Taylor expansion of $f(x)$ in ascending powers of $\left(x - \frac{1}{2}\right)$.

   c Hence find the first four terms of this Taylor expansion.

11 Obtain the general solution of the differential equation

$$x\frac{dy}{dx} + 2y = \cos x, \quad x > 0$$

giving your answer in the form $y = f(x)$.   [(c) Edexcel Limited 2007]

12 The equation of a curve C satisfies the differential equation

$$\frac{d^2y}{dx^2} + y^2 = x^3, \text{ where } y = -2 \text{ and } \frac{dy}{dx} = 3 \text{ at } x = 0.$$

   a Find the value of $\dfrac{d^2y}{dx^2}$ when $x = 0$.

   b Show that $\left(\dfrac{d^3y}{dx^3}\right)_0 = 12$

   c Find a series solution of this differential equation, in ascending powers of x, up to and including the term in x^3.

   d Use your answer to **c** to estimate the gradient of the curve C at $x = 0.1$.

13 The curve C has polar equation $r = 6\cos\theta, \quad -\frac{1}{2}\pi \leqslant \theta < \frac{1}{2}\pi,$

   and the line D has polar equation $r = 3\sec\left(\frac{1}{3}\pi - \theta\right), \quad -\frac{1}{6}\pi < \theta < \frac{5}{6}\pi.$

   a Find a cartesian equation of C and a cartesian equation of D.

   b Sketch on the same diagram the graphs of C and D, indicating where each cuts the initial line.

   The graphs of C and D intersect at the points P and Q.

   c Find the polar coordinates of P and Q.   [(c) Edexcel Limited 2005]

FP2

14 $\dfrac{d^2y}{d\theta^2} + 9y = 2\sin 3\theta$

    a Find the value of the constant λ such that $\lambda\theta\cos 3\theta$ is a particular integral of this differential equation.

    b Hence find the general solution of this differential equation.

    A solution curve C of this differential equation passes through the point $\left(0, \dfrac{1}{3}\right)$. The gradient of C at this point is $-\dfrac{1}{3}$.

    c Find the equation of C and the coordinates of the points in the interval $0 \leqslant \theta \leqslant \dfrac{1}{3}\pi$ where C crosses the θ-axis.

15 a Sketch the curve with polar equation
$$r = 3\cos 2\theta, \quad -\frac{1}{4}\pi \leqslant \theta < \frac{1}{4}\pi$$

    b Find the area of the smaller finite region enclosed between the curve and the half-line $\theta = \dfrac{1}{6}\pi$.

    c Find the exact distance between the two tangents which are parallel to the initial line.   [(c) Edexcel Limited 2004]

16 a Show that the substitution $y = vx$ transforms the differential equation
$$x^2\frac{d^2y}{dx^2} + 2x(3x-1)\frac{dy}{dx} + (2 - 6x + 9x^2)y = x^3e^{-2x} \quad (1)$$
into the differential equation
$$\frac{d^2v}{dx^2} + 6\frac{dv}{dx} + 9v = e^{-2x} \quad\quad\quad (2)$$

    b Find the general solution of differential equation (2).

    c Hence write down the general solution of differential equation (1).

17 $(1 + 2x)\dfrac{dy}{dx} = x + 4y^2$

    a Show that
$$(1+2x)\frac{d^2y}{dx^2} = 1 + 2(4y-1)\frac{dy}{dx} \quad (1)$$

    b Differentiate equation (1) with respect to x to obtain an equation involving $\dfrac{d^3y}{dx^3}, \dfrac{d^2y}{dx^2}, \dfrac{dy}{dx}, x$ and y.

    Given that $y = \dfrac{1}{2}$ at $x = 0$,

    c find a series solution for y, in ascending powers of x, up to and including the term in x^3.   [(c) Edexcel Limited 2006]

18

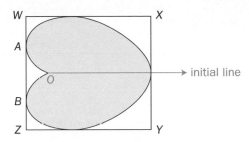

The diagram shows a sketch of the cardioid C with equation

$r - a(1 + \cos 0), \quad -\pi < 0 \leqslant \pi.$

Also shown are the tangents to C that are parallel and perpendicular to the initial line

These tangents form a rectangle $WXYZ$.

a Find the area of the finite region, shaded in the diagram, bounded by the curve C.

b Find the polar coordinates of the points A and B where WZ touches the curve C.

c Hence find the length WX.

Given that the length of WZ is $\frac{3}{2}\sqrt{3}a$,

d find the area of the rectangle $WXYZ$.

A heart-shape is modelled by the cardioid C, where $a = 10$ cm. The heart shape is cut from the rectangular card $WXYZ$, shown in the diagram.

e Find a numerical value for the area of the card wasted in making this heart shape.

[(c) Edexcel Limited 2003]

FP2

19 For the differential equation

$$\left(x + y\right)\frac{d^2y}{dx^2} + x\frac{dy}{dx} - y = 1, \text{ where } y = 0 \text{ and } \frac{dy}{dx} = 2 \text{ at } x = 1.$$

a find the value of $\frac{d^2y}{dx^2}$ at $x = 1$.

b show that $\left(x + y\right)\frac{d^3y}{dx^3} = -\frac{d^2y}{dx^2}\left(1 + x + \frac{dy}{dx}\right)$

c hence find a series solution, in ascending powers of $(x - 1)$, for the given differential equation, up to and including the term in $(x - 1)^3$.

20 The equation of a curve C satisfies the differential equation

$$xy\frac{dy}{dx} = 2y^2 + x^2 \qquad (1)$$

a Show that the substitution $y = ux$ transforms differential equation (1) into the differential equation

$$xu\frac{du}{dx} = u^2 + 1 \qquad (2)$$

b Show that the general solution of equation (2) is given by

$$u^2 = Ax^2 - 1, \quad \text{for } A \text{ an arbitrary constant.}$$

c Hence find the general solution of differential equation (1). Give your answer in the form $y^2 = f(x)$

d Given that C passes through the point $P\left(1, \sqrt{3}\right)$, find the x-coordinates of the points, not at the origin, where C crosses the x-axis.

21 a Show that $y = \frac{1}{2}x^2e^x$ is a solution of the differential equation

$$\frac{d^2y}{dx^2} - 2\frac{dy}{dx} + y = e^x$$

b Solve the differential equation

$$\frac{d^2y}{dx^2} - 2\frac{dy}{dx} + y = e^x$$

given that at $x = 0$, $y = 1$ and $\frac{dy}{dx} = 2$.

[(c) Edexcel Limited 2002]

Answers

Chapter 1

Before you start

1 **a** $-1 < x < 8$ **b** $x < -1, x > \frac{1}{2}$

   c $x \leqslant -\frac{1}{2}, x \geqslant 3$ **d** All $x \in \mathbb{R}$

2 **a** $\frac{7}{(x-3)}$ **b** $\frac{(2x+1)(x+1)}{(x-1)}$ **c** $\frac{(3x+1)(x-1)}{(x+1)(3x-1)}$

3 **a i** **ii**

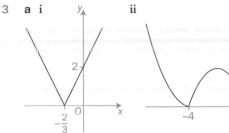

   b

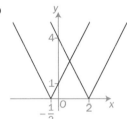

Exercise 1.1

1 **a** $x > 3, x < \frac{1}{2}$ **b** $\frac{1}{3} < x < 2$ **c** $-\frac{5}{4} \leqslant x < \frac{5}{3}$

   d $-\frac{3}{2} < x < 2$ **e** $x \leqslant \frac{3}{5}, x > 3$ **f** $x < \frac{3}{4}, x > \frac{9}{2}$

   g $0 \leqslant x < \frac{3}{2}$

2 **a** $\frac{1}{6} \leqslant x < \frac{3}{2}$ **b** $x < 0, x \geqslant \frac{3}{2}$

3 **a** $x > 3$ **b** $x < \frac{1}{2}$ **c** $x < -\frac{2}{3}$ **d** $x < \frac{3}{2}$

4 **a i** $0 \leqslant x < \frac{3}{2}$ **ii** $x \leqslant \frac{1}{2}, x > 2$

   b $0 \leqslant x \leqslant \frac{1}{2}$

5 **a i** $\frac{2}{3} < x < 2$ **ii** $x \leqslant \frac{3}{5}, x > 3$

6 **a** $2a < x < 4a$ **b** $a = \frac{1}{2}$

7 **a** $\frac{1}{4} \leqslant x \leqslant \frac{1}{3}$ **b** $a = 4$

8 **a** $x < -3, x > 2$ **b** $\frac{1}{2} \leqslant x < 1$ **c** $-1 \leqslant x < 2$

9 **a** $-\frac{1}{2} < x < \frac{5}{2}$

10 **a** $x < -k, x > k$ **b** $\frac{1}{2}k < x < k$

   c $-2k < x < -\frac{1}{2}k$ **d** $2k < x < 4k$

11 **a** $x \leqslant \frac{1}{k^2}, x > \frac{1}{k}$ **b** $x < \frac{1}{h^2}, x > -\frac{1}{h}$

12 **b** e.g. $f(x) = x - 1, g(x) = x - 2$

Exercise 1.2

1 **a** $-2 < x < -1$ **b** $x < -2, x > \frac{1}{2}$

   c $x < 1, x > 6$ **d** $-6 < x \leqslant -4$

   e $x < -\frac{3}{2}, x > \frac{1}{2}$ **f** $-2 \leqslant x < 0$

   g $-\frac{2}{3} < x \leqslant \frac{2}{9}$ **h** $x \leqslant 1, x > \frac{3}{2}$

2 **a** $x > 2$ **b** $x > -\frac{1}{2}$ **c** $x > \frac{1}{3}$

3 Tom has wrongly assumed that
   $3 - x > 0$ for all x. The solution is $x > 3$.

4 **a i** $-2 \leqslant x < \frac{3}{2}$ **ii** $\frac{5}{6} \leqslant x < 2$

   b $\frac{5}{6} \leqslant x < \frac{3}{2}$

5 $x = -42$

6 **a** $-a < x < 2$ **b i** $-a < x < 2$

                  ii $x < -2a, x > 2$

7 **a** $\frac{1}{\sqrt{2}} < x < \sqrt{2}$ **b** $x \leqslant \frac{2\sqrt{2}}{3}, x > \sqrt{2}$

   c $2(\sqrt{2} - 1) \leqslant x < \sqrt{2}$

8 **a** $\frac{1}{2} < x < \frac{4}{3}$ **b** $x < 1$ **c** $x < -3, x \geqslant \frac{3}{11}$

Exercise 1.3

1 **a** $-1 < x < \frac{4}{5}, x > 2$ **b** $x < -\frac{5}{4}, \frac{1}{2} < x < 3$

   c $-2 < x \leqslant -\frac{1}{2}, x > \frac{1}{2}$ **d** $-3 < x \leqslant \frac{1}{3}, x \geqslant \frac{5}{2}$

   e $-3 < x < 0, x > 2$ **f** $x \leqslant -3, -1 \leqslant x < 1, x > 3$

2 **a** $-3 < x < 1, x > 4$ **b** $-4 < x < -2, x > \frac{1}{2}$

   c $x < -1, \frac{1}{2} < x < 2$ **d** $x < -3, -1 \leqslant x < \frac{1}{2}$

   e $x < -3, 2 < x < 3, x > 5$

3 **a** $x < 2$ **b** $-2 < x < 1$ **c** $x < -3, x > 3$

4 **a** $0 < x < 1, x < -2$ **b** $-3 < x < -\frac{3}{2}, x > 0$

   c $-\frac{1}{2} < x \leqslant \frac{5}{4}, x > 3$ **d** $x < -\frac{1}{3}, \frac{1}{4} < x \leqslant 2$

   e $\frac{6}{7} < x < 2, x > 6$ **f** $-3 < x \leqslant -\frac{1}{3}, x > 1$

5 **a** $x < -1, 0 < x < 2$ **b** $-3 < x < \frac{1}{2}, x > 3$

   c $x \leqslant -2, 1 < x \leqslant 2$ **d** $-\frac{3}{2} < x \leqslant -1, x \geqslant \frac{1}{4}$

   e $x < -1, 0 < x < 1$ **f** $-\frac{1}{2} \leqslant x < 1, x \geqslant 3$

   g $x \leqslant -3, -2 < x < 2, x \geqslant 3$
   h $x \leqslant -5, -3 < x \leqslant 2, x > 3$ **i** $x > 0$

6 **a** $-k < x < k$ **b** $x < -k, -1 < x < 1, x > k$

   c $x < -k, x > k$ **d** $0 \leqslant x < \frac{1}{k}$

Exercise 1.4

1 a $x=-2, x=3$ **b** $x=-1, x=\frac{11}{3}$

c $x=-10, x=6$ **d** $x=\frac{1}{2}, x=1$

e $x=\frac{4}{3}, x=6$ **f** $x=-4, x=1$

2 a $x=\pm2$ **b** $x=\pm2, x=\pm4$ **c** $x=-2, x=6$

3 a

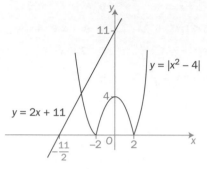

b $x=-3, x=5$

4 a $x=-1, x=4$ **b** $x=1, x=9$ **c** $x=-1, x=\frac{2}{3}$

d $x=-\frac{3}{2}, x=1, x=2$ **e** $x=\pm1, x=3$

5 a $x=\pm2, x=\pm2\sqrt{2}$ **b** $x=1, x=1\pm\sqrt{2}$

c $x=1+\sqrt{2}, x=3+\sqrt{2}$

d $x=-2, x=-\frac{1}{8}(1+\sqrt{33})$

6 a $x=-\frac{1}{3}, x=3$ **b** $x=-2, x=\frac{1}{3}$ **c** $x=-1$

7 a $x=-2, x=6, x=-2\pm\sqrt{10}$

b $x=-1, x=5, x=-2\pm\sqrt{7}$

c $x=\pm\sqrt{2}, x=3\pm\sqrt{7}$

d $x=0, x=1\pm\sqrt{13}$

e $x=\frac{1}{2}(3\pm\sqrt{19})$ **f** $x=\frac{1}{4}(-1\pm\sqrt{5}), \frac{1}{4}(-3\pm\sqrt{29})$

9 $a=8, x=0, x=\pm4$

10 a $x=\pm\sqrt{2k^2+1}$ **b** $x=\pm\frac{1}{\sqrt{2}}k, x=\pm\sqrt{\frac{3}{2}}k$

c $x=0, x=k(1+\sqrt{3})$

Exercise 1.5

1 a $-2<x<3$ **b** $x<-\frac{2}{3}, x>2$

c $-\frac{1}{2}\leqslant x\leqslant-\frac{1}{4}$ **d** $x<\frac{1}{2}, x>3$

e $x\geqslant2$ **f** All $x\in\mathbb{R}$

2 a $x<-3, x>3$ **b** $-3<x<-1, 1<x<3$

c $x\leqslant2, x\geqslant4$ **d** $-4\leqslant x\leqslant0, 1\leqslant x\leqslant3$

e $x<0, 0<x<1, x>3$ **f** $x<0, \frac{3}{2}<x<2, x>\frac{7}{2}$

g $-2<x<0$

3 a $-\sqrt{14}<x<-2, 2<x<\sqrt{14}$

b $x\leqslant-2, -\frac{1}{\sqrt{2}}\leqslant x\leqslant\frac{1}{\sqrt{2}}, x\geqslant2$

c $-1-2\sqrt{2}<x<-1$

d $-2-\sqrt{10}\leqslant x\leqslant-1-\sqrt{3}, -1+\sqrt{3}\leqslant x\leqslant-2+\sqrt{10}$

4 a $x^2+4x-1\equiv(x+2)^2-5$

b Vertex at $(-2,-5)$

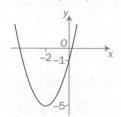

c $-5\leqslant x\leqslant-3, -1\leqslant x\leqslant1$

5 a $-4<x<-1, 0<x<1$

b $x<0, 2-\sqrt{2}<x<2+\sqrt{2}, x>4$

c $x\leqslant-5-3\sqrt{3}, x\geqslant-5+3\sqrt{3}$

6 a $x<-2, x>0$ **b** $x<-\frac{1}{2}$

c $x<-3, -2<x<2, x>3$

d $-2-\sqrt{3}\leqslant x\leqslant-1-\sqrt{2}, -2+\sqrt{3}\leqslant x\leqslant-1+\sqrt{2}$

e $x\leqslant2-\sqrt{2}, 4-2\sqrt{2}\leqslant x\leqslant2+\sqrt{2}, x\geqslant4+2\sqrt{2}$

7 $0<x\leqslant1$

8 a

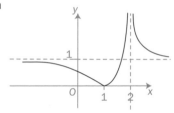

b i $x\leqslant\frac{5}{3}, x\geqslant3$ **ii** $\frac{3}{2}<x<2, x>2$

iii $\frac{1}{2}(3-\sqrt{5})<x<\frac{1}{2}(1+\sqrt{5}), x>\frac{1}{2}(3+\sqrt{5})$

Review 1

1 a

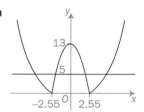

b $-3\leqslant x\leqslant-2, 2\leqslant x\leqslant3$

2 $\frac{1}{2}<x<3$ **3** $x<-1, x>0$ **4** $x=1, 2, 3$

5 a The graphs intersect when $x=\frac{1}{4}$ and $x=1$.

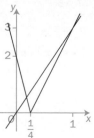

b $x\leqslant\frac{1}{4}, x\geqslant1$

6 a

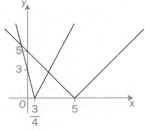

b $x < -\frac{2}{3}, x > \frac{8}{5}$

7 $-1 + 2\sqrt{2} < x < 1 + \sqrt{2}$

8 a

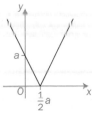

b $-\frac{1}{3}a < x < 3a$

9 $x < -1, \frac{2}{3} < x < 4$ **10** $-\frac{1}{2}k < x < k$

11 a

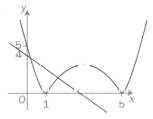

b $x = 2 - \sqrt{3}, x = 4 - \sqrt{7}$

c $x \leqslant 2 - \sqrt{3}, x \geqslant 4 - \sqrt{7}$

12 $0 < x < \frac{2}{3}, 1 < x < 2$ **13** $x \leqslant 1, x \geqslant 3$

14 a

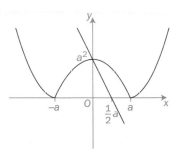

b $-a(\sqrt{3} + 1) < x < 0$

15 a $x = \pm a, x = 2a$

b

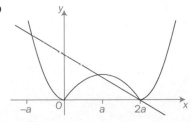

c $-a \leqslant x \leqslant a, x = 2a$

16 a $x = -2$

b

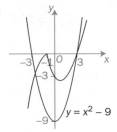

$y = x^2 - 9$

c $x < -2$ or $x > 3$

17 a

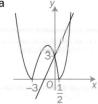

b $x = \frac{1}{2}(-1 + \sqrt{13}), x = 0$

c $x < 0, x > \frac{1}{2}(-1 + \sqrt{13})$

Chapter 2

Before you start

1 **a** $n(n + 1)(n + 3)$ **b** $n(n - 1)(n + 3)$
   c $n(n + 1)(n + 2)$ **d** $(4n - 1)(2n + 3)$

2 **a** $n(n + 4)$ **b** $n^2(2n + 3)$
   c $(n - 1)n(n + 1)(n + 2)$

3 **a** $\dfrac{2n + 3}{n + 1}$ **b** $\dfrac{n^2 + n + 1}{n(n + 1)}$ **c** $\dfrac{(n + 4)(n - 4)}{(n + 2)(n + 8)}$

4 **a** $\dfrac{1}{(r + 1)} + \dfrac{2}{(r - 2)}$ **b** $\dfrac{1}{2(r - 1)} - \dfrac{3}{2(r + 1)}$

   c $\dfrac{2}{r} - \dfrac{4}{(r + 1)} + \dfrac{2}{(r + 2)}$

Exercise 2.1

1 a -20 **b** 13   **c** -78 **d** 48

2 a $r(r + 1) - r(r - 1) \equiv 2r$

3 a $r^2 - (r - 1)^2 \equiv 2r - 1$ **c** 300

4 a $A = 8$ **5 c** $\dfrac{3}{55}$

6 a $\dfrac{1}{(3r - 2)} - \dfrac{1}{(3r + 1)}$ **b** $\dfrac{3n}{3n + 1}$

7 a $\dfrac{1}{(3r - 5)} - \dfrac{1}{(3r + 1)}$ **c** 0.0630 (3 s.f.)

8 b $\ln(1^3 + 2^3 + \cdots + n^3)$ or $2\ln(1 + 2 + 3 + \cdots + n)$

9 a $\dfrac{1}{r} - \dfrac{2}{(r + 1)} + \dfrac{1}{(r + 2)}$

10 a $A = 1, B = 1, C = -1$ **c** $\dfrac{n(2n - 1)(n + 2)}{(2n + 1)(n - 1)}$

11 a $A = 3, B = 3$

Review 2

1 a $(r + 2)^2 - r^2 = 4r + 4$

2 a $\dfrac{1}{(4r - 1)} - \dfrac{1}{(4r + 3)}$ **c** 0.00206 (3 s.f.)

3 a $A = 6, B = 12, C = 8$

4 a $\dfrac{2}{(r-1)} - \dfrac{2}{(r+1)}$

c i $P=1, Q=2$ **ii** $\dfrac{(n-1)(n+2)(2n+1)}{2n(n+1)}$

5 b $1 - \dfrac{1}{(n+1)^2}, A=1, B=-1$ **c** 0.06

6 a $\dfrac{1}{(4r-3)} - \dfrac{1}{(4r+1)}$ **7 a** $A=64, B=16$

8 a $\dfrac{\frac{1}{4}}{(2r-1)} - \dfrac{\frac{1}{4}}{(2r+3)}$

9 b $\dfrac{n(4n+5)}{(n+1)}$ **c** 36.1

10 b $\dfrac{(n-1)(n^2+2n+2)}{2n}$ **11 a** $\dfrac{1}{(r+1)} - \dfrac{2}{r} + \dfrac{1}{(r-1)}$

Chapter 3

Before you start

1 a $z = 4\sqrt{2}\left(\cos\left(-\tfrac{1}{4}\pi\right) + i\sin\left(-\tfrac{1}{4}\pi\right)\right)$

b $z = 2\left(\cos\left(\tfrac{2}{3}\pi\right) + i\sin\left(\tfrac{2}{3}\pi\right)\right)$

c $z = 4\left(\cos\left(-\tfrac{5}{6}\pi\right) + i\sin\left(-\tfrac{5}{6}\pi\right)\right)$

d $z = \tfrac{2}{3}\left(\cos\left(\tfrac{1}{3}\pi\right) + i\sin\left(\tfrac{1}{3}\pi\right)\right)$

2 a i $y = 7.5$ **ii** $6y + 8x = 39$
   iii $10y + 4x = 9$
   b Centre $(6,0)$, radius $= 5$
3 a $a^4 + 4a^3b + 6a^2b^2 + 4ab^3 + b^4$
   b $p^3 + 6p^2q + 12pq^2 + 8q^3$
   c $a^5 - 5a^4b + 10a^3b^2 - 10a^2b^3 + 5ab^4 - b^5$

Exercise 3.1

1 a i **b** -1 **c** $\dfrac{\sqrt{3}}{2} + \dfrac{1}{2}i$

d $\dfrac{1}{\sqrt{2}} - \dfrac{1}{\sqrt{2}}i$ **e** $\dfrac{1}{2} - \dfrac{\sqrt{3}}{2}i$ **f** $-\dfrac{1}{2} + \dfrac{\sqrt{3}}{2}i$

2 a $3\left(\cos\left(\tfrac{2}{7}\pi\right) + i\sin\left(\tfrac{2}{7}\pi\right)\right)$

b $\tfrac{1}{2}\left(\cos\left(-\tfrac{1}{5}\pi\right) + i\sin\left(-\tfrac{1}{5}\pi\right)\right)$

c $4(\cos 2 + i\sin 2)$

3 a $2e^{\frac{\pi}{3}i}$ **b** $4e^{-\frac{\pi}{6}i}$ **c** $4\sqrt{2}e^{\frac{3\pi}{4}i}$

d $2\sqrt{2}e^{-\frac{5\pi}{6}i}$ **e** $5e^{0i}$ **f** $8e^{-\frac{\pi}{2}i}$

4 a -1 **b** $\dfrac{\sqrt{3}}{2} + \dfrac{1}{2}i$ **c** $6i$ **d** $\dfrac{3\sqrt{3}}{8} + \dfrac{3}{8}i$

5 a $\cos 5\theta + i\sin 5\theta$ **b** $2(\cos 4\theta + i\sin 4\theta)$
   c $\cos 2\theta + i\sin 2\theta$ **d** $8(\cos 3\theta + i\sin 3\theta)$

6 a $\arg(z) = \tfrac{2}{3}\pi$ **b** $\arg(w) = \tfrac{1}{3}\pi$

7 $\arg(w) = -\tfrac{1}{4}\pi, \tfrac{3}{4}\pi$

8 a $z^2 = r^2e^{2\theta i}$ **b** $z^* = re^{-i\theta}$ **c** $\dfrac{1}{z} = \dfrac{1}{r}e^{-i\theta}$

9 a $3\sqrt{2}e^{\left(\frac{1}{4}\pi + 2\pi k\right)i}$ **b** $2\sqrt{5}e^{\left(-\frac{1}{6}\pi + 2\pi k\right)i}$ **c** $4e^{\left(\frac{2}{3}\pi + 2\pi k\right)i}$

11 a $iz = e^{i\left(\theta + \frac{1}{2}\pi\right)}$ **b** $z + iz = \sqrt{2}e^{i\left(\theta + \frac{1}{4}\pi\right)}$

c $1 + z^2 = (2\cos\theta)\,e^{i\theta}$ **d** $1 - z^2 = (2\sin\theta)e^{i\left(\theta - \frac{\pi}{2}\right)}$

12 a i **c** $e^{-\pi(1+4k)}, k \in \mathbb{Z}$

13 a $y = 1$ **c** $y = e^{i\theta}$

d Proof valid provided one first shows that the usual rules of calculus apply over the complex numbers.

Exercise 3.2

1 a $z = 3, 3e^{\frac{2\pi}{3}i}, 3e^{\frac{4\pi}{3}i}$ **b** $z = e^{\frac{\pi}{4}i}, e^{\frac{3\pi}{4}i}, e^{\frac{5\pi}{4}i}, e^{\frac{7\pi}{4}i}$

c $z = 2e^{\frac{\pi}{10}i}, 2e^{\frac{\pi}{2}i}, 2e^{\frac{9\pi}{10}i}, 2e^{\frac{13\pi}{10}i}, 2e^{\frac{17\pi}{10}i}$

d $z = \dfrac{1}{\sqrt{2}}e^{0i}, \dfrac{1}{\sqrt{2}}e^{\frac{\pi}{3}i}, \dfrac{1}{\sqrt{2}}e^{\frac{2\pi}{3}i}, \dfrac{1}{\sqrt{2}}e^{\pi i}, \dfrac{1}{\sqrt{2}}e^{\frac{4\pi}{3}i}, \dfrac{1}{\sqrt{2}}e^{\frac{5\pi}{3}i}$

2 a $z = \sqrt{2}e^{\frac{\pi}{12}i}, \sqrt{2}e^{\frac{3\pi}{4}i}, \sqrt{2}e^{-\frac{7\pi}{12}i}$

b $z = 2e^{-\frac{5\pi}{12}i}, 2e^{-\frac{11\pi}{12}i}, 2e^{\frac{\pi}{12}i}, 2e^{\frac{7\pi}{12}i}$

c $z = \dfrac{1}{\sqrt{2}}e^{\frac{5\pi}{12}i}, \dfrac{1}{\sqrt{2}}e^{-\frac{\pi}{4}i}, \dfrac{1}{\sqrt{2}}e^{-\frac{11\pi}{12}i}$

d $z = 2^{\frac{1}{4}}e^{\frac{3\pi}{8}i}, 2^{\frac{1}{4}}e^{-\frac{5\pi}{8}i}$

3 a $z = \dfrac{1}{2}, -\dfrac{1}{4} + \dfrac{\sqrt{3}}{4}i, -\dfrac{1}{4} - \dfrac{\sqrt{3}}{4}i$

b

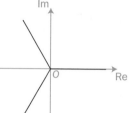

c When these three vectors are added nose-to-tail, they form the sides of an equilateral triangle. The resultant sum points to the zero complex number.

4 a $\pm\sqrt{3}, \pm\sqrt{3}i$

b

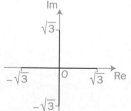

c Area $= 6$

5 a $z = 2, 2e^{\frac{\pi}{3}i}, 2e^{\frac{2\pi}{3}i}, -2, 2e^{-\frac{\pi}{3}i}, 2e^{-\frac{2\pi}{3}i}$

b

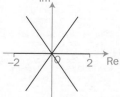

6 a $z = \sqrt{2}\left(\cos\left(\frac{\pi}{9}\right) + i\sin\left(\frac{\pi}{9}\right)\right),$

$\sqrt{2}\left(\cos\left(\frac{7\pi}{9}\right) + i\sin\left(\frac{7\pi}{9}\right)\right),$

$\sqrt{2}\left(\cos\left(-\frac{5\pi}{9}\right) + i\sin\left(-\frac{5\pi}{9}\right)\right)$

b

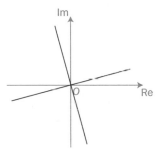

c Area $= \frac{3}{2}\sqrt{3}$

7 a -1 **b** 1 **c** -1 **d** 0

8 a 2 **b** -1 **c** -1

9 a $z = e^{\frac{\pi}{12}i}, e^{\frac{7\pi}{12}i}, e^{-\frac{5\pi}{12}i}, e^{-\frac{11\pi}{12}i}$

b

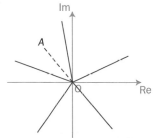

c 0 or $\sqrt{2}$

10 a $z = 2^{-\frac{1}{2}}e^{\frac{3\pi}{20}i}, 2^{-\frac{1}{2}}e^{\frac{11\pi}{20}i}, 2^{-\frac{1}{2}}e^{\frac{19\pi}{20}i}, 2^{-\frac{1}{2}}e^{-\frac{\pi}{4}i}, 2^{-\frac{1}{2}}e^{-\frac{13\pi}{20}i}$

b

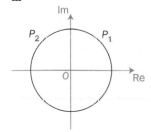

c $w = -4a$, $\lambda = -4$

11 a $n = 5$ **b** $\cos\left(\frac{2\pi}{5}\right) + i\sin\left(\frac{2\pi}{5}\right)$

12 a $z = \sqrt{2}e^{\frac{7\pi}{12}i}, \sqrt{2}e^{\frac{11}{12}i}$ **b** $-\frac{\sqrt{3}}{2} + i\frac{\sqrt{3}}{2}$

14 a $\arg w = \frac{2\pi}{n}$

$n > 4 \therefore \frac{2\pi}{n} < \frac{1}{2}\pi$. Hence w lies in the 1st quadrant.

d $\arg(w^2 - 1) = \left(\frac{n+4}{2n}\right)\pi$

Exercise 3.3

1 a -1 **b** i **c** $-i$

d $8 + 8\sqrt{3}i$ **e** $-\frac{1}{2} + \frac{\sqrt{3}}{2}i$ **f** $\frac{1}{4} - \frac{1}{4}i$

2 a $\frac{1}{2} - \frac{\sqrt{3}}{2}i$ **b** $-\frac{1}{2} + \frac{\sqrt{3}}{2}i$

3 a i **b** $2\sqrt{2} + 2\sqrt{2}i$ **c** -2 **d** 1

4 a 4096 **b** $512i$

c $-32 - 32\sqrt{3}i$ **d** $-\frac{\sqrt{3}}{2} + \frac{1}{2}i$

5 a $\sqrt{2}\left(\cos\left(\frac{\pi}{4}\right) + i\sin\left(\frac{\pi}{4}\right)\right)$

6 b $\lambda - \frac{\sqrt{2}}{\lambda}$ **c** $a^{-3} - \frac{1}{2\sqrt{2}}$

8 a $\sin\theta$

10 a $z^2 = r^2(\cos 2\theta + i\sin 2\theta)$

b 2nd quadrant

c A stretch, scale factor r, and an anticlockwise rotation through θ^c about the origin.

e i e.g. **ii** e.g.

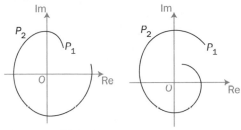

iii

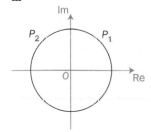

Exercise 3.4

2 b $\theta = \frac{1}{6}\pi, \frac{5}{6}\pi$ **3 b** $\theta = 0, \pi, 2\pi$

4 a $A = \frac{1}{8}, B = -\frac{1}{2}, C = \frac{3}{8}$ **b** $\frac{1}{32}\sin 4\theta - \frac{1}{4}\sin 2\theta + \frac{3}{8}\theta + c$

5 b $\frac{203}{30}$

6 a $\sin 5\theta - 5\sin 3\theta + 10\sin\theta$

b $\theta = 0, \frac{1}{3}\pi, \frac{2}{3}\pi, \pi, \frac{4}{3}\pi, \frac{5}{3}\pi, 2\pi$

7 b $2\cos 3\theta + 6\cos\theta$ **c** $\frac{2}{3}\sin 3\theta + 6\sin\theta + c$

9 b $\theta = \frac{1}{3}\pi, \frac{1}{2}\pi, \frac{2}{3}\pi$ **c** $2\sin 6\theta$

11 c $\frac{4\sqrt{2}}{7}, q = \frac{4}{7}$ **12 b** $16\sin^4\theta - 12\sin^2\theta + 1$

13 a $-2\cos 6\theta + 12\cos 4\theta - 30\cos 2\theta + 20$

c $-\frac{1}{32}\cos 12\theta + \frac{3}{16}\cos 8\theta - \frac{15}{32}\cos 4\theta + \frac{5}{16}$

FP2

Exercise 3.5

1 a $(x-4)^2 + y^2 = 9$
  Circle, centre $(4,0)$, radius 3
b $x^2 + (y+6)^2 = 36$
  Circle, centre $(-6,0)$, radius 6
c $(x-3)^2 + (y+1)^2 = 1$
  Circle, centre $(3,-1)$, radius 1
d $(x-2)^2 + (y-4)^2 = 9$
  Circle, centre $(2,4)$, radius 3
e $(x+4)^2 + (y-1)^2 = 4$
  Circle, centre $(-4,1)$, radius 2
f $(x+3)^2 + (y+2)^2 = 16$
  Circle, centre $(-3,-2)$, radius 4

2 a $y = -2x - 3$
  Straight line
  x-axis intercept $\left(-\frac{3}{2}\right), 0$
  y-axis intercept $(0,-3)$
  Gradient -2
b $y = -1$
  Straight, horizontal line
  y-axis intercept $(0,-1)$
c $3y = 4 - x$
  Straight line
  x-axis intercept $(4,0)$
  y-axis intercept $\left(0, \frac{4}{3}\right)$
  Gradient $-\frac{1}{3}$
d $x = -\frac{7}{2}$
  Straight vertical line
  x-axis intercept $\left(-\frac{7}{2}, 0\right)$
e $y = -3x - 7$
  Straight line
  x-axis intercept $\left(-\frac{7}{3}, 0\right)$
  y-axis intercept $(0,-7)$
  Gradient -3
f $5y = 7x - 8$
  Straight line
  x-axis intercept $\left(\frac{8}{7}, 0\right)$
  y-axis intercept $\left(0, -\frac{8}{5}\right)$
  Gradient $\frac{7}{5}$

3 a
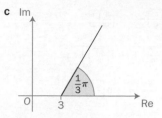

b
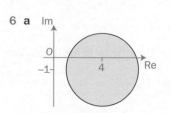

c
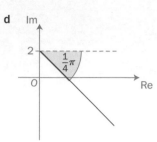

d
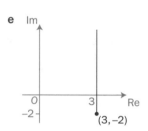

e
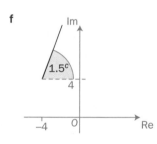

f
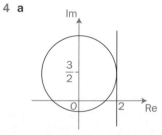

4 a
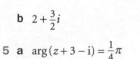

b $2 + \frac{3}{2}i$

5 a $\arg(z+3-i) = \frac{1}{4}\pi$ **b** $|z-2-4i| = 2$

c $\arg(z-\sqrt{3}-3i) = \frac{5}{6}\pi$ **d** $|z-4| = |z-6i|$

6 a

FP2

b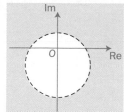

c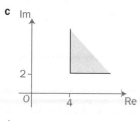

d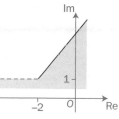

e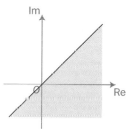

f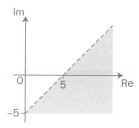

7 a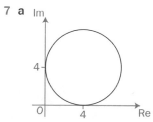

b $4(\sqrt{2}+1)$

8 a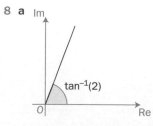

b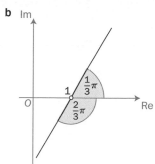

c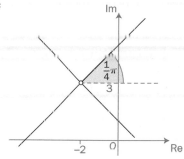

9 a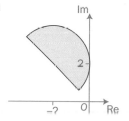

b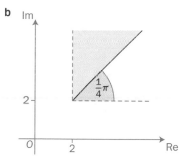

c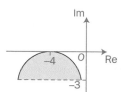

d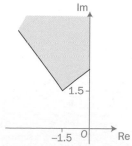

10 a

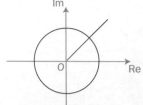

b $z = \sqrt{8} + i\sqrt{8}$

11 a $(x-1)^2 + y^2 = 25$

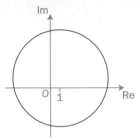

b $0 \pm 2\sqrt{6}i$

c $y = 2 - x$

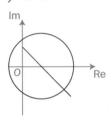

d $z = 5 - 3i$

12 a

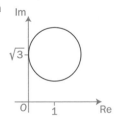

b Minimum arg $z = \frac{1}{6}\pi$

13 b $a = 3, b = -1, c = 3$

14 a $P: y = -2x - 2, \ Q: (x+5)^2 + (y-8)^2 = 5$

b $z = -4 + 6i, -6 + 10i$

d

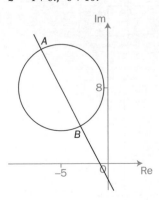

15 a

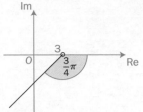

b

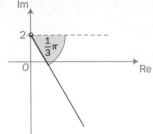

c

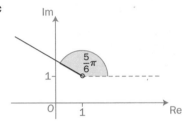

d

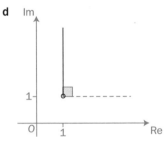

16 a

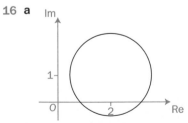

c $z = \frac{1}{2} + \left(1 + \frac{\sqrt{3}}{2}\right)i$

17 a

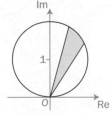

b $-\frac{\sqrt{3}}{4} + \frac{1}{12}\pi + \frac{1}{2}$

Exercise 3.6

1 a

b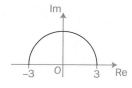

c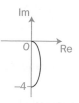

d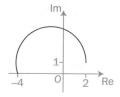

e

f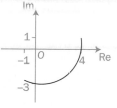

2 a $(x-7)^2 + y^2 = 4$

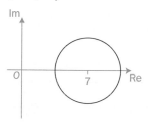

b $(x-1)^2 + (y+1)^2 = 9$

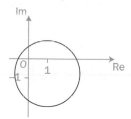

c $(x-1)^2 + y^2 = 8$

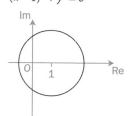

d $(x+1)^2 + (y-1.5)^2 = 11.25$

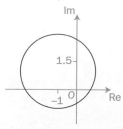

e $(x+1)^2 + (y-2)^2 = 16$

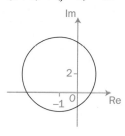

f $(x-5)^2 + (y-1)^2 = 12$

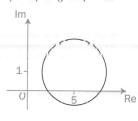

3 b

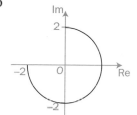

c $0 + 0i$

Three points corresponding to complex numbers $-2, 2i$ and $-2i$ lie on C and are equidistant from O.

4 a

b A semi-circle. Centre $(3,4)$ and radius $2\sqrt{2}$

c $5 + 2\sqrt{2}, a = 5, b = 2, c = 2$

5 a $|z - (6 - 3i)| = \sqrt{5}\,|z - (2 + i)|,\ a = 6 - 3i,\ b = 2 + i$

b $(x-1)^2 + (y-2)^2 = 10$

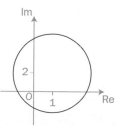

6 a $r = \frac{1}{2}$ **b** $\frac{1}{6}\pi$ **c** $\tan(\arg(z-1)) = -\frac{1}{\sqrt{3}}$

7 b i

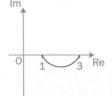

ii

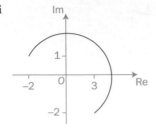

8 a

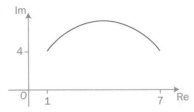

b The perpendicular bisector of the chord joining $(1,4)$ and $(7,4)$ passes through the centre of this circle. Hence the centre lies on the line $x = 4$.

d Radius $2\sqrt{3}$, centre $(4, 4 - \sqrt{3})$

9 $r = k|c|$

10 a

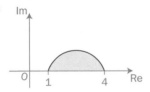

b

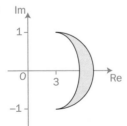

11 a i $\sqrt{3}i$ **ii** $-\sqrt{3}$

b

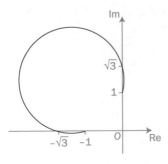

c $\left(\dfrac{-1 - \sqrt{3}}{2}, \dfrac{1 + \sqrt{3}}{2} \right)$

12 a Centre $(6, -3)$, radius 6 **c** $6 \pm 3\sqrt{3}$

   d ii $a = 6 + 3\sqrt{3}$, $b = 6 - 3\sqrt{3}$, $\theta = \dfrac{2}{3}\pi$

13 b

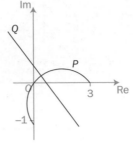

c $|v| = 1$, $\arg v = \dfrac{1}{2}\pi$ **d** $z = 1 + i$

14 b

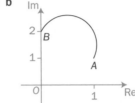

d $|z| = 1 + \sqrt{3}$, $\arg z = \dfrac{1}{3}\pi$

Exercise 3.7

1 a $|w - (2 + i)| = 1$: Circle, centre $(2, 1)$, radius 1

   b $|w - (3 - i)| = 1$: Circle, centre $(3, -1)$, radius 1

   c $\left| w - \dfrac{1}{2} \right| = \dfrac{1}{2}$: Circle, centre $\left(\dfrac{1}{2}, 0 \right)$, radius $\dfrac{1}{2}$

   d $|w - (-1)| = 1$: Circle, centre $(-1, 0)$, radius 1

   e $|w - (-1 + i)| = \sqrt{2}$: Circle, centre $(-1, 1)$, radius $\sqrt{2}$

   f $|w - (-i)| = |w|$: Perpendicular bisector of the line joining $(0, 0)$ and $(0, -1)$

2 a i $|w| = 16$: Circle, centre O, radius 16

   ii $\arg(w) = \dfrac{1}{2}\pi$: Half-line from O at $\dfrac{1}{2}\pi$ radians to the horizontal.

   b $r = \dfrac{1}{2}$

3 a $\arg(w + 1) = \dfrac{\pi}{4}$

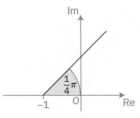

   b $\arg(w + 3) = \dfrac{\pi}{4}$

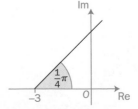

FP2

c $\arg(w + 1 + 2i) = \frac{3\pi}{4}$

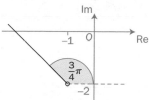

d $\arg(w) = -\frac{\pi}{4}$

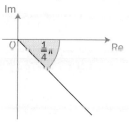

4 a $|w| = |w - 1|$

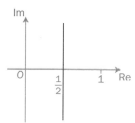

b $\left|w - \left(1 - \frac{1}{2}i\right)\right| = |w - 1|$

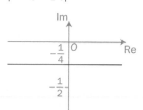

c $\left|w - \frac{1}{5}(-2 + 4i)\right| = \frac{1}{\sqrt{5}}$

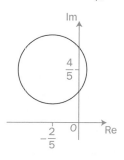

d $\arg\left(\frac{w - i}{w + 3}\right) = \frac{3}{4}\pi$

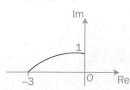

6 The image is the line $u = 1$, excluding the point corresponding to $1 + 0i$

7 a The line is given by $\left|w - \frac{3}{4}\right| = \left|w - \frac{3}{2}\right|$

b

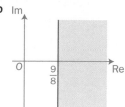

8 a

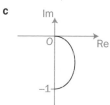

b Centre $(2, -1)$, radius 2

9 a

b

c

10 a $u = \frac{x(2x - 1)}{x^2 + (x - 1)^2}$, $v = \frac{x}{x^2 + (x - 1)^2}$ **11 c** $\sqrt{2}$

12 b

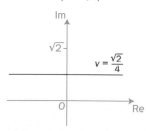

c $a = \frac{1}{2} + \frac{1}{2}i$

d $|w - \sqrt{2}i| = \left|w + \frac{i}{\sqrt{2}}\right|$

e $\text{Im}(a) = \frac{\sqrt{2}}{4}$

FP2

Review 3

1 $z = \sqrt{3}e^{-\frac{1}{8}\pi i}, \sqrt{3}e^{\frac{3}{8}\pi i}, \sqrt{3}e^{\frac{7}{8}\pi i}, \sqrt{3}e^{-\frac{5}{8}\pi i}$

2 a $z = \sqrt{2}e^{-\frac{i\pi}{20}}, \sqrt{2}e^{\frac{i7\pi}{20}}, \sqrt{2}e^{\frac{i3\pi}{4}}, \sqrt{2}e^{-\frac{i9\pi}{20}}, \sqrt{2}e^{-\frac{i17\pi}{20}}$

b

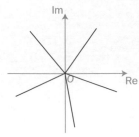

3 a $z = \frac{3}{2}(\sqrt{3} - i), 3i, -\frac{3}{2}(\sqrt{3} + i)$ **b** $27i$ **c** 0

4 a $z^9 = (-4)^9, a = -4$ **b** $-\dfrac{4095}{64}$

5 b $16\cos^4\theta \equiv 2\cos 4\theta + 8\cos 2\theta + 6$,
$a = 2, b = 8, c = 6$

   c $\cos\theta = \pm\dfrac{1}{\sqrt{8}}$

6 a $\sin 5\theta \equiv 5\cos^4\theta\sin\theta - 10\cos^2\theta\sin^3\theta + \sin^5\theta$
$a = 5, b = -10, c = 1$

   c $\theta = \pm\dfrac{1}{6}\pi$

7 c Maximum

8 a $\sin 6\theta \equiv 6\cos^5\theta\sin\theta - 20\cos^3\theta\sin^3\theta + 6\cos\theta\sin^5\theta$

   c $\tan 6\theta = -\dfrac{10\sqrt{2}}{23}$

9 a $64\sin^6\theta \equiv -2\cos 6\theta + 12\cos 4\theta - 30\cos 2\theta + 20$

   c Minimum value $= \dfrac{1}{4}$ when $\theta = \dfrac{1}{4}\pi$

10 a i $(x-2)^2 + (y-2)^2 = 2$ **ii** $y = 2 - x, x < 2$
   b $1 + i$
   c

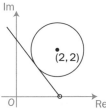

The half-line is a tangent to the circle.

11 a $y = 2x + \dfrac{1}{2}$

   b c

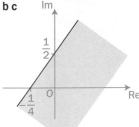

   d $r = \dfrac{9\sqrt{5}}{10}$

12 b

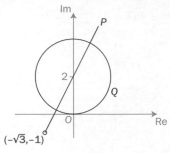

$(-\sqrt{3}, -1)$

   c $1 + (\sqrt{3} + 2)i, -1 + (2 - \sqrt{3})i$

13 a Centre $(2, 1)$, radius 2 **c** $|z_0| = 1$

14 b

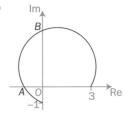

   c Centre $(1, 1)$, radius $r = \sqrt{5}$

15 a

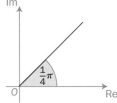

16 a

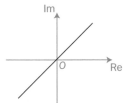

   b $\left(u - \dfrac{1}{2}\right)^2 + \left(v - \dfrac{1}{2}\right)^2 = \dfrac{1}{2}$

17 a

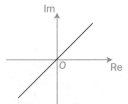

   b $(u + 2)^2 + (v - 2)^2 = 4$

   c

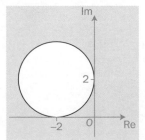

Revision 1

1 a

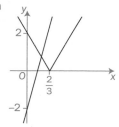

b $x < \dfrac{4}{7}$

2 b $\dfrac{20}{861}$

3 a $z = 2\left(\cos\left(\dfrac{\pi}{12}\right) + i\sin\left(\dfrac{\pi}{12}\right)\right),\ 2\left(\cos\left(\dfrac{3\pi}{4}\right) + i\sin\left(\dfrac{3\pi}{4}\right)\right),$
$2\left(\cos\left(-\dfrac{7\pi}{12}\right) + i\sin\left(-\dfrac{7\pi}{12}\right)\right)$

b

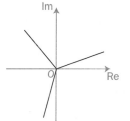

4 a

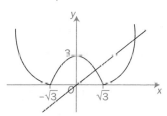

b $\sqrt{7} - 2 < x < 2 + \sqrt{7}$

5 a $\dfrac{1}{(2r+1)} - \dfrac{1}{(2r+3)}$ **c** $\dfrac{1}{81}$

6 c $\theta = 0, \dfrac{1}{4}\pi, \dfrac{3}{4}\pi, \pi, \dfrac{5}{4}\pi, \dfrac{7}{4}\pi$

7 $x \leqslant -1, x \geqslant \dfrac{2}{3}$

8 a Centre $(4,1)$, radius 2

c $|z - (4 + i)| = 2, u = 4 + i, r = 2$

9 a $\dfrac{2r-1}{r^2(r-1)^2}$ **10** $x < 0, 2 < x < 4$

11 b i $\operatorname{Re}(z) = -\dfrac{1}{2}$ **12 a** $A = 24, B = 2$

13 a $v = u - 1$

c

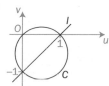

14 $-9 \leqslant x \leqslant -2, -\dfrac{3}{2} < x < 1$

15 a $\dfrac{1}{(2r-3)} - \dfrac{1}{(2r+1)}$

c $\dfrac{n(4n^2 - 2n + 1)}{(4n^2 + 1)(2n+1)(2n-1)}$

16 a

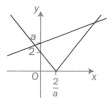

b $\dfrac{2-a}{a+2} < x < \dfrac{a+2}{a-2}$

17 b $\dfrac{3}{2}\pi^2$

18 a

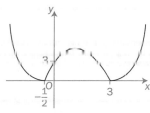

b $-1 < x < 2 - \sqrt{5}$

19 b $-1 + 0i$

c

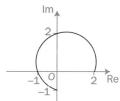

d i $\dfrac{1}{2} + \dfrac{1}{2}i$

20 a $4r^3$

21 a

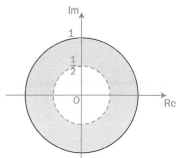

b Centre $(2,0)$, radius 3 **c** $u = -\dfrac{7}{4}$

d

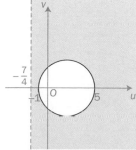

22 a $\dfrac{1}{(r+2)} - \dfrac{1}{(r+3)}$

c i $\dfrac{n(n+2)^2}{(n+3)}$ **ii** 432.25

FP2

Chapter 4

Before you start

1   a $y = \dfrac{1}{c - x^2}$ b $y = \left(\dfrac{3}{2}\ln|x| + c\right)^{\frac{2}{3}}$

    c $y = \pm\sqrt{x + \ln|x| + c}$

2   a $y = \sec\theta$

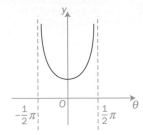

    b $y = x$

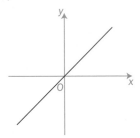

    c $y = \dfrac{2}{x}$

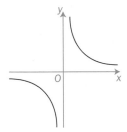

Exercise 4.1

1   a $y = \sqrt[3]{2x + \dfrac{1}{x} + c}$ b $y = Ae^{t^2 + t}$

    c $y = (2\sin\theta + c)^2$

2   a $y^2 = \ln(2x^2 + 1) + 4$ b $\ln(1 - y^3) = \dfrac{3}{t} - 3$

    c $\ln(y + 1) = 2\ln(\sqrt{x} + 1) + \ln 2$

3   b $y = 4\ln x$

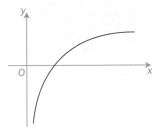

4   a $\dfrac{1}{2}\ln(x^2 + 1) + c$ b $y = A\sqrt{x^2 + 1}$

    c   ii   As $x \to \infty$, $y \to x^+$

5   b $y = Ae^{\tan\theta - \theta}$

6   b $y^2 = 4xe^{2x} - e^{2x} + c$

7   a $\dfrac{1}{(x-1)} - \dfrac{1}{(x+1)}$ b $\dfrac{1}{4}$

    c

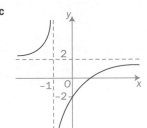

8   $y = \arctan\left(\sqrt{t^2 - 1}\right)$

9   d $y = \dfrac{1}{4}(x + e)$ e $x = \dfrac{1}{e}$

Exercise 4.2

1   a $\dfrac{dm}{dt} = -km$, $k > 0$ a constant

    b $m = 20.5e^{-0.00206t}$

2   a $\dfrac{dh}{dt} = kh$ where $k = 0.025$

    c   Approximately 2 minutes

3   a $\dfrac{dV}{dt} = k(500\pi - V)$, $k > 0$ a constant.

    c   i   3 minutes 19 seconds
        ii   109 cm³/min

4   a $\dfrac{dA}{dt} = \dfrac{kt^3}{(25 - A)}$ c   9 km²

5   a   i $\dfrac{dA}{dt} = \dfrac{k\sqrt{\pi}}{\sqrt{A}}$, $k > 0$ a constant.

        ii   As $t \to \infty$, $A \to \infty$. The model is unrealistic in the long run.

    b   i $\dfrac{dA}{dt} = \dfrac{kA^2}{\pi^2}$, $k > 0$ a constant.

        ii   When $t \geqslant 6$, $21 - 4t < 0$ giving $A < 0$, which is not possible.

6   a $\dfrac{dN}{dt} = kN(150 - N)$ c $\lambda = 0.5$ (1 d.p.)

    e   i   80 sheep   ii   18.7 sheep per day
    f   As $t \to \infty$, $N \to 150$, suggesting that, unless treated, all the sheep will eventually become infected. This seems realistic.

Exercise 4.3

1   a $y = \dfrac{x^3 - 3x + c}{x + 1}$ b $y = \dfrac{x + c}{x^2}$

    c $y = \dfrac{x - 2\ln|x| + c}{(x - 2)}$ d $y = \dfrac{x^3 + 3x^2 + c}{(x + 2)^2}$

    e $y = x^2 + \dfrac{1}{2x} + cx$ f $y = \dfrac{\sin(\theta^2) + c}{2\theta^2}$

    g $x = \dfrac{t^3 + c}{3\sqrt{t}}$ h $y = \dfrac{2x^3 + 3x^2 + c}{xe^x}$

    i $y = x\left(1 + ce^{-\frac{1}{2}x^2}\right)$

2 b $y = e^{-x}$

c

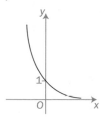

3 a i $y = \dfrac{1}{2} + ce^{-2x}$ **ii** $y = \dfrac{1}{2}(1 - Ae^{-2x^2})$

b The two expressions appear to be different but in fact describe the same family of solution curves, since both c and A are arbitrary constants.

4 a $\dfrac{1}{x} - \dfrac{1}{(x+1)}$ **b** $y = \left(\dfrac{x+1}{x}\right)(\ln|x+1| + c)$

5 a $x\sin x + \cos x + c$ **c** $y = \dfrac{x\sin x + \cos x + 1}{\cos x}$

6 b $y = \dfrac{2\sin^3\theta + c}{\sin\theta}$

d

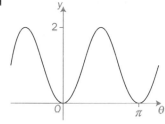

7 a $-\dfrac{1}{(x+1)} + c$ **b** $y = \left(\dfrac{x+1}{e^x}\right)\left(c - \dfrac{1}{(x+1)}\right)$

8 $y = \dfrac{e^x}{\sin x}(e^x + c)$

9 b i $y = \dfrac{x+c}{\tan x}$ **ii** $y = \dfrac{2x - \tan x + c}{\tan x}$

Exercise 4.4

1 a i $\dfrac{du}{dx} = 2xu^2, u = -\dfrac{1}{(x^2+c)}$ **ii** $y = -\dfrac{x}{(x^2+c)}$

b i $u\dfrac{du}{dx} = 3x, u^2 = 3x^2 + c$ **ii** $y = x\sqrt{3x^2 + c}$

c i $u^2\dfrac{du}{dx} = \dfrac{1}{x^2}, u^3 = -\dfrac{3}{x} + c$ **ii** $y = x\left(c - \dfrac{3}{x}\right)^{\frac{1}{3}}$

d i $\dfrac{du}{dx} = \dfrac{u^2}{x}, u = \dfrac{-1}{\ln|x| + c}$ **ii** $\dfrac{-x}{\ln|x| + c}$

e i $\dfrac{du}{dx} = \dfrac{\sqrt{u}}{\sqrt{x}}, u^{\frac{1}{2}} = x^{\frac{1}{2}} + c$ **ii** $y = x(\sqrt{x} + c)^2$

2 b $u = \dfrac{1}{e^x + c}$

c $y = xe^{-x}$

3 b $y = x\left(Ae^{\frac{1}{2}x^2} - 1\right)^{\frac{1}{2}}$

4 a $y^2 = x^2(\sin x + c)$ **b** $y^2 = x^2 \ln|\sin x|$

c $y^2 = \dfrac{x^2}{c - 2\tan x}$

5 b $u = \sqrt{c - \cos 2x}$ **c ii** $\left(\dfrac{3}{4}\pi, \dfrac{3}{4}\pi\right)$

6 c $v = \dfrac{1}{c - x^2}$

7 a $\dfrac{1}{(u-2)} - \dfrac{1}{(u+2)}$ **d** $y = \dfrac{2x(1-x)}{(1+x)}$

8 b $u = 2 \pm \dfrac{1}{x}\sqrt{3x^2 + A}$ **d i** $y = 2x \pm \sqrt{3x^2 + 1}$

9 a $\displaystyle\int \dfrac{1}{x\ln x}\,dx - \ln|\ln|x|| + c$ **c** $u = e^{Au}$, for $A \in \mathbb{R}$

d i $y = xe^{-2x}$

ii

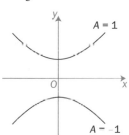

Review 4

1 a $y = A(x^2 + 1)$

b e.g.

2 a $te^t - e^t + c$

3 a $\dfrac{dr}{dt} = \dfrac{100k}{\pi r^2\left(1 - \frac{1}{10}t\right)}, k > 0$, a constant.

c 5.7 hours   **d** $t = 10 \Rightarrow h = 0$ so r is undefined.

4 a $y = \dfrac{1}{2}x^3 + cx$

b $y = \dfrac{1}{2}x(x^2 - 1)$

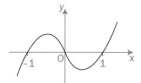

5 a $x - \ln|x+1| + c$

6 b ii $x = 2\cos\theta(\sin 2\theta + 2\theta + c)$

7 b $u = \pm\sqrt{Ax^2 + 1}$ **c** $y^2 = x^2(Ax^2 + 1)$

8 a $\dfrac{dv}{dx} = 2v(1 - x), v = Ae^{2x - x^2}$

c $y = 3ex$ **d** $x = \dfrac{1}{2}(1 + \sqrt{3})$

FP2

Chapter 5

Before you start

1 a $x = \frac{1}{2} \pm \frac{1}{2}i$ b $x = -\frac{1}{3}, x = -1$

   c $m = -\frac{1}{2} \pm \frac{3}{2}i$ d $m = -\frac{3}{2}$

2 a $3e^x + 16e^{2x}$ b $4(\cos 2x - x\sin 2x)$

   c $4e^{2x}(2x + 3)$

   d $4(\cos 2x - \sin 2x - x(\sin 2x + \cos 2x))$

3 a 0   b 4

Exercise 5.1

1 a $y = Ae^{5x} + Be^{-x}$ b $y = Ae^{-2x} + Be^{-3x}$

   c $y = e^{2x}(Ax + B)$ d $y = Ae^{2x} + Be^{-\frac{1}{2}x}$

   e $y = e^{2x}(A\cos x + B\sin x)$

   f $y = e^{-4x}(Ax + B)$ g $y = e^{\frac{1}{2}x}(Ax + B)$

   h $y = e^{-\frac{2}{5}x}\left(A\cos\left(\frac{1}{5}x\right) + B\sin\left(\frac{1}{5}x\right)\right)$

   i $y = e^{-\frac{3}{5}x}\left(A\cos\left(\frac{4}{5}x\right) + B\sin\left(\frac{4}{5}x\right)\right)$

2 a $u = Ae^{-3x} + Be^{-4x}$ b $x = e^{-\frac{3}{2}t}(At + B)$

   c $y = A\cos 3x + B\sin 3x$

3 a $y = A + Be^{-\frac{2}{3}x}$ b $y \to A$, a constant, as $x \to \infty$

4 a $y = e^{-3x}(Ax + B)$ b $x = e^{-3t}(A\cos 2t + B\sin 2t)$

   c $y = Ae^t + Be^{-\frac{1}{4}t}$ d $y = A\cos\left(\frac{3}{2}x\right) + B\sin\left(\frac{3}{2}x\right)$

   e $u = Ae^{\frac{2}{3}x} + Be^{-2x}$ f $v = e^{\frac{1}{3}t}(At + B)$

5 b $y = e^{3x}(Ax + B)$ 6 b $e^{2\pi}$

7 a $y = e^{5x}(Ax + B)$ b $\frac{d^2y}{dx^2} + 4\frac{dy}{dx} + 4y = 0$

8 a $\frac{d^2y}{dx^2} = -2e^x \sin x$ c $y = e^x(A\cos x + B\sin x)$

9 b i $x_0 = -\frac{1}{k} - \frac{B}{A}$

    ii The stationary point is a maximum if $y_0 > 0$ and a minimum if $y_0 < 0$.

Exercise 5.2

1 a $y = \frac{1}{3}e^x$ b $y = \frac{1}{4}e^{-3x}$

   c $y = \cos 2x - 3\sin 2x$ d $y = \frac{1}{3}\cos 2x + \frac{2}{3}\sin 2x$

   e $y = -\frac{3}{2}x + \frac{1}{4}$ f $y = -\frac{1}{2}x^2 - \frac{1}{4}x - \frac{7}{16}$

2 a $y = e^{2x}(Ax + B) + \frac{1}{3}e^{-x}$

   b $y = Ae^{2t} + Be^{-t} - \frac{9}{2}t + \frac{9}{4}$

   c $y = e^{-\frac{1}{2}t}(At + B) + \cos t + \sin t$

   d $y = e^{2x}(A\cos x + B\sin x) + 3e^{3x}$

   e $x = e^t(A\cos 3t + B\sin 3t) + \frac{1}{2}t^2 + t - 2$

   f $y = e^{-\theta}(A\cos 2\theta + B\sin 2\theta) + \frac{1}{5}\cos 4\theta + \frac{2}{5}\sin 4\theta$

3 a $y = e^{-4x}(Ax + B) + 7e^{-3x}$ b $y \to 0$ as $x \to \infty$

4 a $y = e^{-2x}(Ax + B) + \sin 2x$

   b

5 b $y = Ae^{2x} + Be^{-3x} + 2xe^{2x}$

6 b $y = Ae^{3x} + B - \frac{1}{2}x^2$

7 a $k = -1$ b $y = A\cos 2x + B\sin 2x - x\cos 2x$

8 a $y = Ae^{-\frac{1}{2}x} + Be^{-2x} + 3x - 12$ b $y \approx 3x - 12$

9 a $y = Ae^{-2\theta} + Be^{-3\theta} - \frac{2}{5}\cos\theta + \frac{2}{5}\sin\theta$

   b $y \approx \frac{2\sqrt{2}}{5}\sin\left(\theta - \frac{1}{4}\pi\right)$

10 a $y = Ae^{-2x} + B - \frac{3}{2}xe^{-2x}$

    $y \to B$, a constant, as $x \to \infty$

   b $y = Ae^{-2x} + B - \frac{1}{6}x^3 + \frac{1}{4}x^2 + \frac{1}{2}x$

    $x \to -\infty$, a constant, as $x \to \infty$

11 b i 15

12 b $y \approx (\cos x + \sin x)^2 \equiv 1 + \sin 2x$

   c $g(x) \approx \sqrt{2}\cos\left(x - \frac{1}{4}\pi\right)$

Exercise 5.3

1 a $y = e^{5x} + 3e^{2x}$ b $y = e^{\frac{3}{2}x}(4 - 3x)$

   c $y = e^{-\frac{2}{3}x}(2x + 5)$ d $y = e^{-2t}(2\cos 3t - \sin 3t)$

   e $x = e^{2\theta}(2\cos 2\theta - \sin 2\theta)$

2 a $y = -5e^x + 4e^{-\frac{1}{2}x} - x^2 + 2x - 7$

   b $y = 3\cos\frac{1}{2}x + 2e^x$

   c $y = e^{-3\theta}(3\theta + 2) - \frac{1}{2}\cos\theta - \frac{1}{2}\sin\theta$

3 b $-\frac{13}{4}$

4 a $y = e^x(\cos 2x - \sin 2x) + \frac{1}{2}\cos x + \frac{1}{2}\sin x$

   b $e^\pi - \frac{1}{2}$

5 a $y = e^{\frac{1}{2}x}\left(1 - \frac{1}{4}x\right) + x - \frac{1}{4}x^2$ b $x = 4$

6 b Gradient $= -1 - \frac{1}{2}\sqrt{2}e^{\frac{3}{4}\pi}$

7 b $y = 2\cos 4t + \left(1 - \frac{\pi}{16}\right)\sin 4t + \frac{1}{2}t\sin 4t$

8 **b** $P(-\ln 2, 0)$
   c ii

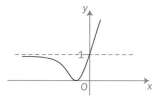

Exercise 5.4

1 **a** $y = x(Ae^{2x} + Be^{-2x})$ **b** $y = x(A + Be^{-x})$

   c $y = xe^{2x}(A\cos 2x + B\sin 2x)$ **d** $y = xe^{-\frac{3}{2}x}(Ax + B)$

2 **b** $y = xe^{-x}(Ax + B)$

3 **b** $u = A\cos\frac{1}{2}x + B\sin\frac{1}{2}x + x^2$

4 **b** $v = e^{\frac{1}{2}x}(Ax + B)$

   c i $y = xe^{-\frac{1}{2}x}(1 - x)$

   ii

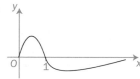

5 **b** $v = Ae^{2x} + B - 6e^x$ **d** $x = \ln 4$

6 **a** $\dfrac{dy}{dx} = \dfrac{dv}{dx} - 1, \dfrac{d^2y}{dx^2} = \dfrac{d^2v}{dx^2}$ **b** $y = e^{\frac{1}{2}x}(Ax + B) - x$

7 **b** $u = A\cos 2x + B\sin 2x - 1$
        $y = x(A\cos 2x + B\sin 2x - 1)$

   c $y = -\dfrac{3}{2}\pi$

   d

8 **a** $\dfrac{d^2v}{dx^2} = x\dfrac{d^2y}{dx^2} + 2\dfrac{dy}{dx}$

   c $v = Ae^{-x} + Be^{-2x}$

        $y = \dfrac{Ae^{-x} + Be^{-2x}}{x}$

9 **a** $\dfrac{d^2y}{dt^2} = -\dfrac{1}{x^2}\dfrac{d^2x}{dt^2} + \dfrac{2}{x^3}\left(\dfrac{dx}{dt}\right)^2$
   c $y = Ae^{2t} + B$ **d** i $x = \dfrac{4}{e^{2t} + 3}$

Review 5

1 **a** $y = Ae^{4x} + Be^{-5x}$ **b** $y = e^{-\frac{1}{3}x}(Ax + B)$
   c $x = e^{4t}(A\cos t + B\sin t)$

2 **a** $y = Ae^{\frac{2}{3}x} + Be^{-3x} - \dfrac{1}{4}e^{-2x}$ **b** ii $y \to -\infty$ as $x \to \infty$

3 **a** $k = \dfrac{1}{2}$ **b** $y = e^{-2x}(Ax + B) + \dfrac{1}{2}x^2e^{-2x}$

   c ii

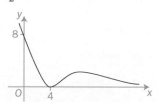

4 **a** $y = Ae^{-\theta} + Be^{-3\theta} + 3\cos\theta$
   b $y = 2e^{-\theta} + e^{-3\theta} + 3\cos\theta$

5 **a** $x = Ae^t + Be^{-3t} - t + \dfrac{2}{3}$

   b $x = Ae^{2t} + B + \dfrac{1}{4}t(3t - 11)$

6 **b** $u = Ae^{\frac{3}{2}x} + Be^x + 4e^{-x}$

7 **b** $v = A\cos 4x + B\sin 4x - \sin 2x$

   c ii $\left(\dfrac{1}{6}\pi, 0\right), \left(\dfrac{5}{6}\pi, 0\right), (0, 0), \left(\dfrac{\pi}{2}, 0\right), (\pi, 0)$

8 **a** $\dfrac{d^2v}{dx^2} = x\dfrac{d^2y}{dx^2} + 2\dfrac{dy}{dx}$ **c** $v = Ae^{-x} + Be^{-\frac{3}{2}x} + 2x - \dfrac{7}{3}$

   d i $y = \dfrac{Ae^{-x} + Be^{-\frac{3}{2}x} + 2x - \dfrac{7}{3}}{x}$ II $y \approx 2$

9 **a** $y = Ae^{\frac{1}{3}t} + Be^{-t} - t^2 - 3t$

   b $y = 3e^{\frac{1}{3}t} + 2e^{-t} - t^2 - 3t$ **c** $y = 2e^3 + 3e^{-1}$

10 **b** $y = 2(1 + x)\sin 3x$ **c** $\left(\pm\dfrac{1}{3}\pi, 0\right), (-1, 0)$

11 **b** $y = e^{-\frac{1}{2}x} - 2e^{-x} + 2xe^{-x}$ **c** $y = -e^{-\frac{1}{2}}$

12 **b** $v = A\cos 4x + B\sin 4x + \dfrac{1}{2}x^2$

   c $y = x\left(A\cos 4x + B\sin 4x + \dfrac{1}{2}x^2\right)$

Chapter 6

Before you start

1 **a** $f'(x) = \sin 2x, f''(x) = 2\cos 2x$

   b $f'(x) = \dfrac{2}{(1 + 2x)}, f''(x) = -\dfrac{4}{(1 + 2x)^2}$

   c $f'(x) = x(1 + 2\ln x), f''(x) = 3 + 2\ln x$
   d $f'(x) = -\tan x, f''(x) = -\sec^2 x$

2 **a** i $1 - 6x + 27x^2 - 108x^3 + \cdots$

     ii $1 - \dfrac{1}{2}x + \dfrac{1}{8}x^2 - \dfrac{1}{16}x^3 + \cdots$

     iii $-1 + 4x - 15x^2 + 60x^3 - \cdots$

   b ii $\sqrt{3} \approx 1.7349\cdots = 1.73\,(2\,\text{d.p.})$

3 **a** $5y^4\dfrac{dy}{dx}$ **b** $2x^2y\dfrac{dy}{dx} + 2xy^2$

   c $x^3\dfrac{d^2y}{dx^2} + 3x^2\dfrac{dy}{dx}$ **d** $y^2\dfrac{d^2y}{dx^2} + 2y\left(\dfrac{dy}{dx}\right)^2$

Exercise 6.1

1 **a** $f^{(1)}(x) = 2e^{2x+1}, f^{(2)}(x) = 4e^{2x+1}, f^{(3)}(x) = 8e^{2x+1}$
   b $f^{(1)}(x) = 3\sin(1 - 3x), f^{(2)}(x) = -9\cos(1 - 3x),$
        $f^{(3)}(x) = -27\sin(1 - 3x)$
   c $f^{(1)}(x) = (x + 3)^{-1}, f^{(2)}(x) = -(x + 3)^{-2},$
        $f^{(3)}(x) = 2(x + 3)^{-3}$
   d $f^{(1)}(x) = (2x + 1)^{-\frac{1}{2}}, f^{(2)}(x) = -(2x + 1)^{-\frac{3}{2}},$

        $f^{(3)}(x) = 3(2x + 1)^{-\frac{5}{2}}$

   e $f^{(1)}(x) = e^{2x}(2x + 1), f^{(2)}(x) = 4e^{2x}(x + 1),$
        $f^{(3)}(x) = 4e^{2x}(2x + 3)$
   f $f^{(1)}(x) = x(1 + 2\ln x), f^{(2)}(x) = 3 + 2\ln x, f^{(3)}(x) = \dfrac{2}{x}$

2 a 2   **b** 2   **c** –1

3 a 2   **b** –2   **c** e   **d** 6e

4 b $\left(\dfrac{d^2 y}{dx^2}\right)_1 = -\dfrac{4}{27}, \left(\dfrac{d^3 y}{dx^3}\right)_{-1} = 24$

6 b $N = 3$ **7 b** $2n$

Exercise 6.2

1 a $1 + x + \dfrac{1}{2}x^2 + \dfrac{1}{6}x^3 + \cdots$

  b $x - \dfrac{1}{6}x^3 + \cdots$ **c** $x - \dfrac{1}{2}x^2 + \dfrac{1}{3}x^3 - \cdots$

  d $1 + \dfrac{1}{2}x - \dfrac{1}{8}x^2 + \dfrac{1}{16}x^3 - \cdots$

2 a $\dfrac{1}{2} + \dfrac{\sqrt{3}}{2}x - \dfrac{1}{4}x^2 - \dfrac{\sqrt{3}}{12}x^3 + \cdots$

  b $\dfrac{1}{2} - \sqrt{3}x - x^2 + \dfrac{2\sqrt{3}}{3}x^3 + \cdots$

  c $1 + \dfrac{1}{e}x - \dfrac{1}{2e^2}x^2 + \dfrac{1}{3e^3}x^3 - \cdots$

3 a $1 - 2x + 2x^2 - \cdots$, valid for all x

  b $1 - 8x^2 + \dfrac{32}{3}x^4 - \cdots$, valid for all x

  c $x^2 - \dfrac{1}{2}x^4 + \dfrac{1}{3}x^6 - \cdots$, valid for $-1 \leqslant x \leqslant 1$

  d $1 + \dfrac{3}{2}x + \dfrac{9}{8}x^2 + \dfrac{9}{16}x^3 + \cdots$, valid for all x

4 a $3x - \dfrac{9}{2}x^3 + \dfrac{81}{40}x^5 - \cdots$

5 a $1 + 2x - 2x^2 - \dfrac{4}{3}x^3 - \cdots$, valid for all x

  b $1 + 2x + 3x^2 + \dfrac{10}{3}x^3 + \cdots$, valid for all x

  c $2x + 2x^2 + \dfrac{2}{3}x^3 + \cdots$, valid for $-\dfrac{1}{2} < x \leqslant \dfrac{1}{2}$

  d $1 - 2x + \dfrac{5}{2}x^2 - \dfrac{8}{3}x^3 + \cdots$, valid for $-1 < x < 1$

6 b $\dfrac{1}{\sqrt[3]{e}} \approx \dfrac{58}{81} = 0.71605$ (5 d.p.)

7 a $2x + 2x^2$ **b** $\alpha \approx -\dfrac{1}{6}$

8 a $2x^2 - x^3$ **b** $1 + 2x - 6x^2 - \dfrac{44}{3}x^3 + \cdots$

  c $-x - x^2 - \dfrac{17}{24}x^3 - \cdots$

9 b $-1 + x^2 + \dfrac{2}{3}x^3 + \cdots$

10 a $e^{ax+b} = e^b + ae^b x + \dfrac{a^2 e^b}{2}x^2 + \cdots$

  b $a = 3, k = 2, b = \ln 2$

11 a The full expansion is valid for $-1 \leqslant x < 1$

  b $\ln 2$

12 a $1 + kx^2 + \cdots$ **c** $\alpha \approx 0.2928\cdots = 0.3 (1 \text{ d.p.})$

13 b $f(x) = x - \dfrac{1}{2}x^2 + \dfrac{1}{6}x^3 + \cdots$

  c $\ln(1 - \sin x) = -x - \dfrac{1}{2}x^2 - \dfrac{1}{6}x^3 + \cdots$

14 c Sum of series $= 2 - \dfrac{1}{2}\pi$

15 a $1 + x + \dfrac{1}{2}x^2 + \dfrac{1}{6}x^3 + \dfrac{1}{24}x^4 + \dfrac{1}{120}x^5 + \cdots$

Exercise 6.3

1 a $-(x - \pi) + \dfrac{1}{6}(x - \pi)^3 + \cdots$

  b $\left(x - \dfrac{3}{2}\pi\right) - \dfrac{1}{6}\left(x - \dfrac{3}{2}\pi\right)^3 + \cdots$

  c $4 + 8(x - \ln 2) + 8(x - \ln 2)^2 + \dfrac{16}{3}(x - \ln 2)^3 + \cdots$

2 a $\dfrac{dy}{dx} = \dfrac{1}{x}, \dfrac{d^2 y}{dx^2} = -\dfrac{1}{x^2}, \dfrac{d^3 y}{dx^3} = \dfrac{2}{x^3}$

  b $(x - 1) - \dfrac{1}{2}(x - 1)^2 + \dfrac{1}{3}(x - 1)^3 + \cdots$

  c i use $x = 1.5$: $\ln 1.5 \approx \dfrac{5}{12} = 0.417 (3 \text{ d.p.})$

    ii use $x = 1.2$: $\ln 6 - \ln 5 \approx \dfrac{137}{750} = 0.183 (3 \text{ d.p.})$

3 a $p = 2, q = -\dfrac{4}{3}$ **b** $(x - \pi) - \dfrac{2}{3}(x - \pi)^3 + \cdots$

4 b $1 - \left(x - \dfrac{1}{2}\pi\right)^2 + \dfrac{1}{3}\left(x - \dfrac{1}{2}\pi\right)^4 + \cdots$

5 b $\cos 86° \approx 0.0707\cdots = 0.07$ (2 d.p.)

6 c 0.003   **7 a** $\dfrac{1}{2} - \dfrac{\sqrt{3}}{2}x - \dfrac{1}{4}x^2 + \cdots$

8 a $a = 2$ **b** $\dfrac{1}{4}x - \dfrac{1}{4}x^2 + \dfrac{3}{16}x^3 + \cdots$

9 b $1 + (\ln 2)(x - 1) + \dfrac{\ln 2(\ln 2 - 1)}{2}(x - 1)^2 + \cdots$

10 b $A = -4, B = \dfrac{32}{3}$ **c** $-\dfrac{128}{15}\left(x - \dfrac{1}{4}\pi\right)^5$

11 b $0.09292\cdots = 0.093$ (3 d.p.)

Exercise 6.4

1 a $y = 1 - x + \dfrac{1}{2}x^2 + \dfrac{1}{6}x^3 + \cdots$

  b $y = -2 + 4x - 8x^2 + \dfrac{49}{3}x^3 + \cdots$

  c $y = -1 - x - x^2 - 2x^3 + \cdots$

2 a $y = \dfrac{1}{2} + x - \dfrac{1}{4}x^2 + \dfrac{1}{2}x^3 + \cdots$ **b** $y \approx 0.598$

3 a $y = 1 + 3(x - 1) + 5(x - 1)^2 + \dfrac{17}{3}(x - 1)^3 + \cdots$

  b $y = 2 + 2(x - 1) + \dfrac{4}{3}(x - 1)^2 + \dfrac{2}{3}(x - 1)^3 + \cdots$

4 a $y = -2 + 2x - x^2 - x^3 + \cdots$ **b** $y \approx -1.648$

  c $\left(\dfrac{dy}{dx}\right)_{0.2} \approx 1.47$ (2 d.p.)

5 a $y = 1 + x + \dfrac{1}{2}x^2 + \dfrac{1}{2}x^3 + \cdots$

  b $y = 2 - x + 2x^2 - \dfrac{1}{6}x^3 + \cdots$

  c $y = 2x + \dfrac{1}{2}x^2 - \dfrac{2}{3}x^3 + \cdots$

6 b $\left(\dfrac{d^3 y}{dx^3}\right)_0 = 14$ **c** $y = 3 + x + \dfrac{9}{2}x^2 + \dfrac{7}{3}x^3 + \cdots$

  d $y \approx 3.40$ (2 d.p.)

7 a $x = \pi - t + \dfrac{1}{2}t^2 + \dfrac{1}{6}t^3 + \cdots$

  b $x = 3 - 2t - 2t^2 + \dfrac{5}{6}t^3 + \cdots$

8 a $y_1^{(2)} = 1$. Point P is a minimum.

c $y = 1 + \frac{1}{2}(x-1)^2 - \frac{5}{6}(x-1)^3 + \cdots$

d i $y \approx 1.006$ (3 d.p.) **ii** $y \approx 1.004$ (3 d.p.)

Both values are just greater than 1. This is consistent with the fact that $P(1,1)$ is a minimum.

10 b $\left(\dfrac{d^3 y}{dx^3}\right)_0 = -2$

c $y = 1 - x + \frac{1}{2}x^2 - \frac{1}{3}x^3 + \frac{5}{12}x^4 + \cdots$

11 a $\left(\dfrac{d^3 y}{dx^3}\right)_0 - 5$ **b** $y - 1 + x - \frac{3}{2}x^2 + \frac{5}{6}x^3 + \cdots$

c $h = 1, k = 2$

d $(y(x))^2 = 1 + 2x - 2x^2 - \frac{4}{3}x^3 + \cdots$

Review 6

1 a $1 - 2x^2 + \frac{2}{3}x^4 - \cdots$ **b** $a = 3, b = -3, c = 1$

c $x = \pm\dfrac{1}{\sqrt{2}}$

2 a $f^{(2)}(x) = x(1 - x^2)^{-\frac{3}{2}}$

$f^{(3)}(x) = (1 - x^2)^{-\frac{3}{2}} + 3x^2(1 - x^2)^{-\frac{5}{2}}$

b $x + \frac{1}{6}x^3 + \cdots$ **c** $\dfrac{51}{128}$

3 a i $1 + x + \frac{1}{2}x^2 + \frac{1}{6}x^3 + \cdots$ **ii** $x - \frac{1}{6}x^3 + \cdots$

c $a \approx -0.38196\ldots = -0.38$ (2 d.p.)

4 a $-2\left(x - \frac{1}{4}\pi\right)^2 + \cdots$

5 b $-\dfrac{1}{64}$ **c** $-(x-1) - \frac{1}{2}(x-1)^2 - \frac{1}{3}(x-1)^3 + \cdots$

6 a $\ln 2 + \frac{1}{2}(x-1) - \frac{1}{8}(x-1)^2 + \cdots$

7 a $-3(x - \pi) + \frac{9}{2}(x - \pi)^3 + \cdots$

8 b $\dfrac{3}{2}$ **c** $p = \pm\sqrt{2}$

9 a 4 **c** $\left(\dfrac{d^2 y}{dx^2}\right)_0 = 13$

d $y = 2 + 4x + \frac{13}{2}x^2 + \frac{29}{3}x^3 + \cdots$

10 b -1 **c** $y = 1 + \frac{1}{2}x^2 - \frac{1}{6}x^3 + \cdots$

11 b $\dfrac{d^3 y}{dx^3} = 1 - (x+y)\dfrac{d^2 y}{dx^2} - \dfrac{dy}{dx}\left(1 + \dfrac{dy}{dx}\right) + 2y\dfrac{dy}{dx}$

d $y = 2 - (x-1) + 4(x-1)^2 - \frac{9}{2}(x-1)^3 + \cdots$

12 a $y = 3x - \frac{1}{2}x^2 + \cdots$ **b** $k \approx 0.58$

13 a -2

14 b $k = 2$ **c** -32

Chapter 7

Before you start

1 Area $= \frac{3}{2}, k = \sqrt{17}$

2 a i $-2\sin 4\theta$ **ii** $\frac{1}{2}\sin 4\theta$

b i $\sin 2\theta + 2\theta + c$ **ii** $3\theta - \sin 2\theta + 4\cos\theta + c$

3 a $\theta = 0, \frac{2}{3}\pi$ **b** $\theta = -\pi, 0, \frac{1}{4}\pi, \pi, -\frac{3\pi}{4}$

c $\theta = \pm\frac{1}{2}\pi, \pm\frac{1}{4}\pi$

Exercise 7.1

1 $A\left(4, \frac{1}{6}\pi\right), B\left(3, \frac{19}{24}\pi\right), C\left(4, -\frac{1}{3}\pi\right), D\left(5, -\frac{11}{15}\pi\right)$

2 a $A\left(5, -\frac{1}{2}\pi\right)$ **b** $B\left(3, -\frac{1}{4}\pi\right)$ **c** $C\left(8, \frac{2}{3}\pi\right)$

d $D(4, \pi)$ **e** $E(1, -1.3^c)$ **f** $F(4, 2.3^c)$

g $G\left(\frac{1}{2}, \frac{1}{4}\pi\right)$ **h** $H\left(4, -\frac{1}{2}\pi\right)$

3 b $PQ = \sqrt{13}$

4 a $r = 2$ **c** $BC = 2$

5 a

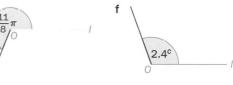

b Arca $= 13.0$ (1 d.p.) **c** $AB = \sqrt{91}$

6 $Q\left(3, -\frac{1}{3}\pi\right)$

Exercise 7.2

1 a

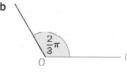

b

c

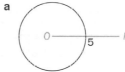

d

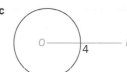

e

f

2 a i ii

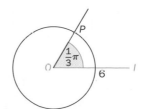

b $P\left(6, \frac{1}{3}\pi\right)$ **c** $Q_1(6,0), Q_2\left(6, \frac{2}{3}\pi\right)$

FP2

3 a

θ	0	$\frac{1}{4}\pi$	$\frac{2}{4}\pi$	$\frac{3}{4}\pi$	π
$r = 3(1 + \cos\theta)$	6	5.1	3	0.9	0

b

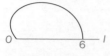

c

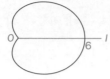

4 a

θ	0	$\frac{1}{8}\pi$	$\frac{2}{8}\pi$	$\frac{3}{8}\pi$	$\frac{4}{8}\pi$
$r = 6\sin\theta$	0	2.3	4.2	5.5	6

b

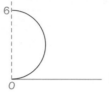

c $\sin(\pi - \theta) \equiv \sin\theta$

d

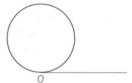

5 a

θ	0	$\frac{1}{4}\pi$	$\frac{2}{4}\pi$	$\frac{3}{4}\pi$	π	$\frac{5}{4}\pi$	$\frac{6}{4}\pi$	$\frac{7}{4}\pi$	2π
$r = \dfrac{10}{1+\theta}$	10	5.6	3.9	3.0	2.4	2.0	1.8	1.5	1.4

b i ii

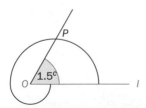

c $P(4, 1.5^c)$

6 a

θ	0	$\frac{1}{4}\pi$	$\frac{1}{2}\pi$
$r = \sqrt{2}\sec\left(\frac{1}{4}\pi - \theta\right)$	2	1.4	2

b

c The points P, Q and R are collinear and Q is the midpoint of PR.

7 b

θ	0	$\frac{1}{16}\pi$	$\frac{2}{16}\pi$	$\frac{3}{16}\pi$	$\frac{4}{16}\pi$
$r = 6\cos 2\theta$	6	5.5	4.2	2.3	0

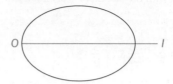

c $P\left(3, \frac{1}{6}\pi\right)$

8 a

θ	0	$\frac{1}{16}\pi$	$\frac{2}{16}\pi$	$\frac{3}{16}\pi$	$\frac{4}{16}\pi$
r	3	2.9	2.5	1.9	0

b

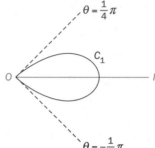

d

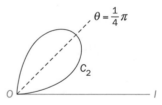

Anti-clockwise rotation, $\frac{1}{6}\pi$ radians about O

e $\theta = \frac{1}{4}\pi$

Exercise 7.3

1 a $x^2 + (y - 2)^2 = 4$
Circle, centre $(0, 2)$, radius 2

b $\left(x - \frac{5}{2}\right)^2 + y^2 = \frac{25}{4}$

Circle, centre $\left(\frac{5}{2}, 0\right)$, radius $\frac{5}{2}$

c $x = \frac{3}{2}$
Straight vertical line through $\left(\frac{3}{2}, 0\right)$

d $y = 4$
Straight horizontal line through $(0, 4)$

e $y = 2 - x$
Straight line
Axes crossings $(0, 2)$ and $(2, 0)$
Gradient -1

f $y = \dfrac{1}{x}$

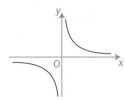

2 **a** $y = \pm\sqrt{16 - x^2}$, $-4 \leqslant x \leqslant 4$

b $y = \pm\sqrt{4 - x^2}$, $-2 \leqslant x \leqslant 2$

c $y = \sqrt{3}x$, $x \geqslant 0$ **d** $y = -\dfrac{1}{\sqrt{3}}x$, $x \geqslant 0$

3 **a** $(x+1)^2 + (y-1)^2 = 2$ **b** $\left(x - \dfrac{1}{2}\right)^2 + \left(y + \dfrac{1}{2}\right)^2 = \dfrac{1}{2}$

c $x^2 - y^2 = 1$ **d** $(x^2 + y^2)^2 = 2xy$

e $x(x^2 + y^2) = y$ **f** $xy(x^2 + y^2) = 3$

4 **b** C has cartesian equation $(x-3)^2 + y^2 = 9$ for $0 \leqslant y \leqslant 3$

c $B\left(3, \dfrac{1}{3}\pi\right)$

5 **a** $r = 4\sec\theta$ **b** $r = \dfrac{1}{2}\operatorname{cosec}\theta$

c $r = 2\cos\theta$ **d** $r = 4\sin\theta$

e $r = 4(\cos\theta + \sin\theta)$ **f** $r = \sqrt{\dfrac{1}{1 + \sin 2\theta}}$

6 **a** C has cartesian equation $x^2 + (y - 2a)^2 = 4a^2$

b **i** $P(0, 2a)$ **ii** $P\left(2a, \dfrac{1}{2}\pi\right)$

c $Q_1\left(2a, \dfrac{1}{6}\pi\right), Q_2\left(2a, \dfrac{5}{6}\pi\right)$

7 **a** $2\cos\left(\dfrac{1}{3}\pi - \theta\right)$

c **i** 3 **ii** $\dfrac{1}{3}\pi$ **iii** $A\left(\dfrac{3}{2}, \dfrac{3\sqrt{3}}{2}\right)$

8 **a** $k = 2$ **c** $B(2\sqrt{7})$, $\tan^{-1}\left(\dfrac{\sqrt{3}}{2}\right)$

Exercise 7.4

3 **a** $P\left(2\sqrt{2}, \dfrac{1}{4}\pi\right), Q\left(2\sqrt{2}, -\dfrac{1}{4}\pi\right)$

b Area = 4

4 **b** $\dfrac{dx}{d\theta} = -\sin\theta + 3\cos^2\theta\sin\theta$

5 **b** $B\left(\dfrac{3}{2}, \dfrac{2}{3}\pi\right)$ **6** **c** Area = 4

7 **a** $\left(1, \dfrac{1}{2}\pi\right), \left(\dfrac{1}{3}, -\dfrac{1}{2}\pi\right)$ **b** $\left(\dfrac{2}{2}, \dfrac{1}{2}\pi\right), \left(\dfrac{2}{2}, \dfrac{5}{2}\pi\right)$

8 **c** Area = $\dfrac{3}{8}\sqrt{3}$, $p = \dfrac{3}{8}$, $q = 3$

Exercise 7.5

1 **a** $\left(\dfrac{\sqrt{3}}{2}, \dfrac{1}{3}\pi\right)$ **b** $\left(\sqrt{3}, \dfrac{1}{2}\pi\right)$

c $\left(5, \dfrac{1}{2}\pi\right)$ **d** $\left(4, \dfrac{\pi}{3}\right), \left(4, \dfrac{5\pi}{3}\right)$

2 **a** $\left(\sqrt{2}, \dfrac{1}{4}\pi\right), \left(\sqrt{2}, \dfrac{3\pi}{4}\right)$

b $(2,0), \left(\dfrac{1}{2}, \dfrac{1}{3}\pi\right), \left(\dfrac{1}{2}, \dfrac{5}{3}\pi\right), (2, 2\pi)$

c $(1,0)$

d $\left(\dfrac{1}{2}, \dfrac{\pi}{6}\right)$

3 **b** $\left(\dfrac{\sqrt{3}}{2}, \dfrac{1}{3}\pi\right)$

4 **a** $P(1,0), Q\left(\dfrac{1}{2}, \dfrac{1}{6}\pi\right)$ **b** Area = $\dfrac{1}{8}$

5 **a** $A\left(2\sqrt{3}, \dfrac{1}{6}\pi\right), B\left(2\sqrt{3}, -\dfrac{1}{6}\pi\right)$

6 **a** $(0,0), \left(\dfrac{3}{2}, \dfrac{1}{2}\pi\right)$

b Figure 3: C_1 passes through $\left(2, \dfrac{1}{2}\pi\right)$, C_2 through $\left(3, \dfrac{1}{2}\pi\right)$ and the graphs intersect at two points only.

7 **b** $(0,0), \left(\dfrac{3}{2}, \dfrac{2}{3}\pi\right)$ **d** $Q\left(\dfrac{1}{4}\sqrt{37}, 2.7^c\right)$

8 **a** $A\left(\dfrac{1}{2}, \dfrac{1}{3}\pi\right), B\left(\dfrac{1}{2}, -\dfrac{1}{3}\pi\right)$

9 **a** $x^2 + (y-2)^2 = 4$, radius = 2

b $A\left(2, \dfrac{1}{6}\pi\right), B\left(2\sqrt{3}, -\dfrac{1}{3}\pi\right)$

d Area = $\dfrac{2}{3}\pi$

Exercise 7.6

1 **a** $\dfrac{1}{4}\pi + \dfrac{1}{2}$ **b** $1 - \dfrac{\sqrt{3}}{2}$

c $\dfrac{1}{8}\pi$ **d** $\dfrac{3}{2}\pi + 4$

2 **b** $\dfrac{1}{12}\pi + \dfrac{1}{8}$

4 **a** 10 **b** $\dfrac{1}{4}\pi$

c $\dfrac{1}{2}(e^{8\pi} - 1)$ **d** $\dfrac{1}{4}\ln 3$

5 **b** $\dfrac{1}{2}a^2(3\pi + 2)$

6 **b** $P\left(2\sqrt{2}, \dfrac{1}{4}\pi\right), Q\left(4, \dfrac{1}{3}\pi\right)$

7 **a** $P\left(2, \dfrac{1}{6}\pi\right), Q\left(2\sqrt{3}, \dfrac{1}{3}\pi\right)$ **b** Area = $\sqrt{3}$

9 **b** 0.86 (2 d.p.)

c **i** The curve is not defined for $\theta > \dfrac{1}{4}\pi$.

Review 7

1 **a** $y = \dfrac{5}{x}$, $x > 0$

b

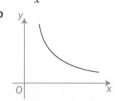

2 a $k = 2\sqrt{2}$, $\alpha = \frac{\pi}{4}$ **b** $y = 4\csc\left(\theta - \frac{1}{4}\pi\right)$

3 a $\left(\frac{3}{2}a, \frac{1}{3}\pi\right)$, $\left(2a, \frac{1}{2}\pi\right)$

4 b $\cos\alpha = \frac{1}{\sqrt{3}}$, $k = 2$ **5 b** $\pi(4 + a^2)$

6 b

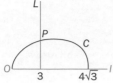

c 3π

7 b $Q\left(1, \frac{2}{3}\pi\right)$ **d** $\pi - \frac{1}{2}\sqrt{3}$

8 a $\frac{3}{4}a^2$

9 a $A\left(\frac{1}{2}a, \frac{1}{6}\pi\right)$

10 b $A\left(\frac{1}{\sqrt{2}}a, \frac{1}{6}\pi\right)$, $B\left(\frac{1}{\sqrt{2}}a, -\frac{1}{6}\pi\right)$

    d $r = \frac{\sqrt{6}}{4}a\sec\theta$, $-\frac{1}{6}\pi \leqslant \theta \leqslant \frac{1}{6}\pi$

Revision 2

1 a $-2\left(x - \frac{1}{4}\pi\right) + \frac{4}{3}\left(x - \frac{1}{4}\pi\right)^3 - \frac{4}{15}\left(x - \frac{1}{4}\pi\right)^5 + \cdots$

    b $-0.416\,147$ (6 d.p.)

2 a $(2,0)$

    b $y = x - 2$, $x \geqslant 2$

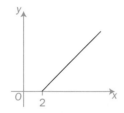

3 $y = \sec x(2\tan x + 3)$

4 a $y = Ae^{-\frac{1}{2}t} + Be^{-3t} + t^2 - t + 1$

    b $y = \frac{4}{5}\left(e^{-\frac{1}{2}t} - e^{-3t}\right) + t^2 - t + 1$

    c $y = \frac{4}{5}\left(e^{-\frac{1}{2}} - e^{-3}\right) + 1 = 1.45$ (3 s.f.)

5 a $x - \frac{1}{2}x^2 + \frac{1}{3}x^3 - \cdots$ **b** $-1 \leqslant x < 1$

    c $2x + \frac{2}{3}x^3 + \cdots$

6 a $P\left(\frac{3}{2}, \frac{1}{6}\pi\right)$

7 b $y = \frac{1}{2} + \frac{1}{4}x + \frac{1}{2}x^2 + \cdots$ **c** $\frac{3}{4}$

8 a $y = e^{-2x}(A\cos x + B\sin x) - 8\cos 2x + \sin 2x$

    b $y \approx \sqrt{65}\sin(2x - 1.45^c)$

9 a $\left(4a, \frac{1}{3}\pi\right)$, $\left(4a, \frac{5}{3}\pi\right)$ **b** $\frac{a^2}{6}(76\pi - 39\sqrt{3})$

c

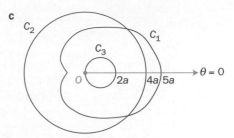

10 b $-\frac{1}{n}$

    c $\ln 2 - \left(x - \frac{1}{2}\right) - \frac{1}{2}\left(x - \frac{1}{2}\right)^2 - \frac{1}{3}\left(x - \frac{1}{2}\right)^3 - \cdots$

11 $y = \frac{x\sin x + \cos x + c}{x^2}$

12 a -4

    c $y = -2 + 3x - 2x^2 + 2x^3 + \cdots$ **d** 2.66

13 a $C: (x - 3)^2 + y^2 = 9$, $D: x + \sqrt{3}y = 6$

    b

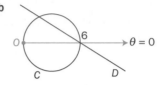

    c $(6, 0)$, $\left(3, \frac{1}{2}\pi\right)$

14 a $\lambda = -\frac{1}{3}$

    b $y = A\cos 3\theta + B\sin 3\theta - \frac{1}{3}\theta\cos 3\theta$

    c $C: y = \frac{1}{3}(1 - \theta)\cos 3\theta$, $\left(\frac{1}{6}\pi, 0\right)$, $(1^c, 0)$

15 a

    b $\frac{3}{32}(2\pi - 3\sqrt{3})$ **c** $\frac{2}{3}\sqrt{6}$

16 b $v = e^{-3x}(Ax + B) + e^{-2x}$

    c $y = x\left(e^{-3x}(Ax + B) + e^{-2x}\right)$

17 b $(1 + 2x)\frac{d^3y}{dx^3} + 2\frac{d^2y}{dx^2} = 8\left(\frac{dy}{dx}\right)^2 + 2(4y - 1)\frac{d^2y}{dx^2}$

    c $y = \frac{1}{2} + x + \frac{3}{2}x^2 + \frac{4}{3}x^3 + \cdots$

18 a $\frac{3}{2}\pi a^2$ **b** $A\left(\frac{1}{2}a, \frac{2}{3}\pi\right)$, $B\left(\frac{1}{2}a, -\frac{2}{3}\pi\right)$

    c $\frac{9}{4}a$ **d** $\frac{27\sqrt{3}}{8}a^2$

    e 113 cm^2 (3 s.f.)

19 a -1 **c** $y = 2(x - 1) - \frac{1}{2}(x - 1)^2 + \frac{2}{3}(x - 1)^3 + \cdots$

20 c $y^2 = x^2(Ax^2 - 1)$ **d** $x = \pm\frac{1}{2}$

21 b $y = \frac{1}{2}e^x(x^2 + 2x + 2)$

FP2

Formulae

The following formulae will be given to you in the exam formulae booklet.
You may also require those formulae listed under Further Pure Mathematics FP1
and Core Mathematics C1–C4.

Area of a sector

$$A = \frac{1}{2}\int r^2 \, d\theta \quad \text{(polar coordinates)}$$

Complex numbers

$$e^{i\theta} = \cos\theta + i\sin\theta$$

$$\{r(\cos\theta + i\sin\theta)\}^n = r^n(\cos n\theta + i\sin n\theta)$$

The roots of $z^n = 1$ are given by $z = e^{\frac{2\pi k i}{n}}$, for $k = 0, 1, 2, \ldots, n-1$

Maclaurin's and Taylor's series

$$f(x) = f(0) + xf'(0) + \frac{x^2}{2!}f''(0) + \cdots + \frac{x^r}{r!}f^{(r)}(0) + \cdots$$

$$f(x) = f(a) + (x-a)f'(a) + \frac{(x-a)^2}{2!}f''(a) + \cdots + \frac{(x-a)^r}{r!}f^{(r)}(a) + \cdots$$

$$f(a+x) = f(a) + xf'(a) + \frac{x^2}{2!}f''(a) + \cdots + \frac{x^r}{r!}f^{(r)}(a) + \cdots$$

$$e^x = \exp(x) = 1 + x + \frac{x^2}{2!} + \cdots + \frac{x^r}{r!} + \cdots \quad \text{for all } x$$

$$\ln(1+x) = x - \frac{x^2}{2} + \frac{x^3}{3} - \cdots + (-1)^{r+1}\frac{x^r}{r} + \cdots \quad (-1 < x \leqslant 1)$$

$$\sin x = x - \frac{x^3}{3!} + \frac{x^5}{5!} - \cdots + (-1)^r \frac{x^{2r+1}}{(2r+1)!} + \cdots \quad \text{for all } x$$

$$\cos x = 1 - \frac{x^2}{2!} + \frac{x^4}{4!} - \cdots + (-1)^r \frac{x^{2r}}{(2r)!} + \cdots \quad \text{for all } x$$

$$\arctan x = x - \frac{x^3}{3} + \frac{x^5}{5} - \cdots + (-1)^r \frac{x^{2r+1}}{2r+1} + \cdots \quad (-1 \leqslant x \leqslant 1)$$

Taylor polynomials

$$f(a+h) = f(a) + hf'(a) + \frac{h^2}{2!}f''(a) + \text{error}$$

$$f(a+h) = f(a) + hf'(a) + \frac{h^2}{2!}f''(a+\xi) \quad (0 < \xi < h)$$

$$f(x) = f(a) + (x-a)f'(a) + \frac{(x-a)^2}{2!}f''(a) + \text{error}$$

$$f(x) = f(a) + (x-a)f'(a) + \frac{(x-a)^2}{2!}f''(\xi) \quad (a < \xi < x)$$

Arbitrary constant A quantity used to represent a real number.

e.g. $\int 3x^2\,dx = x^3 + c$ where c is an arbitrary constant

Area of a sector of a polar curve The area A of the region bounded by the curve $r = f(\theta)$ where $\alpha \leqslant \theta \leqslant \beta$ and the half-lines $\theta = \alpha$, $\theta = \beta$.

Argument, arg(z) An angle θ between the line OP and the positive real axis, where P represents z.

Auxiliary equation The associated quadratic equation for a second-order differential equation.

e.g. $2\dfrac{d^2y}{dx^2} + 3\dfrac{dy}{dx} + y = e^x$ has auxiliary equation

$2m^2 + 3m + 1 = 0$

Binomial expansion The expansion of $(a + bx)^n$ as a power series in ascending powers of x, where $a, b, n \in \mathbb{R}$.

Boundary conditions Known values which are used to find the particular solution of a differential equation.

Cardioid e.g. a polar curve of the form $r = a(1 + \cos\theta)$, where $a > 0$ is a constant.

Chain rule A method for differentiating a function of a function.

e.g. $\dfrac{d}{dx}(e^{2x+1}) = 2e^{2x+1}$

Chord A line whose endpoints are two given points on the circumference of the circle.

Complementary function The general solution of the differential equation $a\dfrac{d^2y}{dx^2} + b\dfrac{dy}{dx} + cy = 0$

Critical values of a rational function The critical values of $f(x) = \dfrac{g(x)}{h(x)}$ are the values of x for which either $g(x) = 0$ or $h(x) = 0$

de Moivre's theorem A general result from which it follows that $(\cos\theta + i\sin\theta)^n \equiv \cos(n\theta) + i\sin(n\theta)$ for any integer n.

Direct proportionality Two quantities P and Q vary in direct proportion (written as $P \propto Q$) if $P = kQ$ for k a constant.

Exponential form of a complex number z A general expression involving $|z|$ and arg(z) e.g. $z = 2i$ has general exponential form $2e^{\frac{1}{2}\pi i(1+4k)}$, $k \in \mathbb{Z}$.

Family of solution curves of a differential equation The collection of all curves whose equations satisfy the differential equation.

First-order differential equation An equation involving a function and its first derivative.

e.g. $\dfrac{dy}{dx} + 3y = 2x + 1$

First-order linear differential equation A differential equation of the form $\dfrac{dy}{dx} + Py = Q$ where P and Q are functions of x.

General solution of a differential equation An expression satisfying the differential equation and which includes one or more arbitrary constants.

Half-line in the complex plane A line, beginning at a fixed point A and which is inclined at a given angle against the positive real axis. The half line does not include point A.

Half-line in the polar coordinate system A line extending from the pole which makes an angle θ with the initial line.

Image of a complex number The result of a applying a transformation to a complex number z.

Implicit differentiation Differentiating a function of one variable with respect to another variable e.g. $\dfrac{d}{dx}(y^2) = 2y\dfrac{dy}{dx}$

Inequality An expression involving one of the symbols $<, \leqslant, >$ or $\geqslant$

Initial line A fixed line from the pole in the polar coordinate system.

Integrating factor The differential equation $\dfrac{dy}{dx} + Py = Q$ has integrating factor $I = e^{\int P\,dx}$

Inverse proportionality Two quantities P and Q are inversely proportional (written as $P \propto \dfrac{1}{Q}$) if $PQ = k$ for k a constant.

Leminscate e.g. a polar curve of the form $r^2 = a^2 \cos 2\theta$, where $a > 0$ is a constant.

Locus (plural Loci) A set of points which are governed by a given rule.

Maclaurin expansion of f(x) The power series $f(x) = f(0) + \dfrac{x}{1!}f^{(1)}(0) + \dfrac{x^2}{2!}f^{(2)}(0) + \cdots$

Method of differences A method for proving results in series.

Modulus-argument form A complex number in the form $z = r(\cos\theta + i\sin\theta)$, where $r = |z|$ and θ is an argument of z.

FP2

Modulus function An expression of the form $|f(x)|$, for $f(x)$ a real-valued function.

Modulus of a complex number The length of the line OP, where point P represents $z = a + bi$; $a, b \in \mathbb{R}$; $|z| = +\sqrt{a^2 + b^2}$

Modulus of a real number The numerical value of a number. e.g. $|-3| = 3$

nth derivative of a function The result of differentiating a function n times, where $n \in \mathbb{Z}^+$.

nth root of unity Any complex number z such that $z^n = 1$, for a given natural number n.

Partial fractions The process of expressing a rational function as the sum of other fractions.

Particular integral A function which satisfies the second-order differential equation $a\dfrac{d^2y}{dx^2} + b\dfrac{dy}{dx} + cy = f(x)$, where $f(x)$ is not the zero function.

Particular solution of a differential equation A function satisfying the differential equation and which does not include an arbitrary constant.

Perpendicular bisector of the line AB The line which is perpendicular to AB and which passes through the midpoint of AB.

Polar coordinates A system in which a point A is specified by its distance r from the pole O and angle θ which the line OA makes with the initial line.

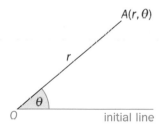

Polar curve A curve defined by the polar equation $r = f(\theta)$

Polar graph The graph of a polar curve.

Pole The fixed point O in the polar coordinate system.

Power series at the point $x = a$ An infinite series of the form $\displaystyle\sum_{r=0}^{\infty} b_r(x - a)^r$ where $b_r \in \mathbb{R}$ for all r.

Product rule $\dfrac{d}{dx}(uv) = u\dfrac{dv}{dx} + v\dfrac{du}{dx}$, where u and v are functions of x.

Rational function A function of the form $f(x) = \dfrac{g(x)}{h(x)}$, where $g(x)$ and $h(x)$ are polynomials.

e.g $f(x) = \dfrac{2x + 1}{x^2 + 1}$

Second-order differential equation An equation of the form $a\dfrac{d^2y}{dx^2} + b\dfrac{dy}{dx} + cy = f(x)$ where a, b and c are constants.

Separating the variables A method for solving a differential equation.

Tangent to a polar curve, parallel to the initial line A horizontal line, or lines, found by solving $\dfrac{dy}{d\theta} = 0$ where $y = f(\theta)\sin\theta$

Tangent to a polar curve, perpendicular to the initial line A vertical line, or lines, found by solving $\dfrac{dx}{d\theta} = 0$ where $x = f(\theta)\cos\theta$

Taylor expansion of $f(x)$ about $x = a$, where a is a constant The power series $f(x) = f(a) + \dfrac{(x - a)}{1!}f^{(1)}(a) + \dfrac{(x - a)^2}{2!}f^{(2)}(a) + \cdots$

Transformation of the complex plane A mapping from the set of complex numbers to itself.

Trial function See Particular Integral.

w-plane The image of the complex plane under a transformation.

FP2

Index